# Calculus:
## Early Transcendental Functions

### FIFTH EDITION

**Ron Larson**

The Pennsylvania State University; The Behrend College

**Bruce Edwards**

University of Florida

BROOKS/COLE
CENGAGE Learning™

Australia • Brazil • Japan • Korea • Mexico • Singapore • Spain • United Kingdom • United States

© 2011 Brooks/Cole, Cengage Learning

ALL RIGHTS RESERVED. No part of this work covered by the copyright herein may be reproduced, transmitted, stored, or used in any form or by any means graphic, electronic, or mechanical, including but not limited to photocopying, recording, scanning, digitizing, taping, Web distribution, information networks, or information storage and retrieval systems, except as permitted under Section 107 or 108 of the 1976 United States Copyright Act, without the prior written permission of the publisher.

For product information and technology assistance, contact us at
**Cengage Learning Customer & Sales Support,
1-800-354-9706**

For permission to use material from this text or product, submit all requests online at **www.cengage.com/permissions**
Further permissions questions can be emailed to
**permissionrequest@cengage.com**

ISBN-13: 978-0-538-73671-8
ISBN-10: 0-538-73671-2

**Brooks/Cole**
20 Channel Center Street
Boston, MA 02210
USA

Cengage Learning is a leading provider of customized learning solutions with office locations around the globe, including Singapore, the United Kingdom, Australia, Mexico, Brazil, and Japan. Locate your local office at:
**www.cengage.com/global**

Cengage Learning products are represented in Canada by Nelson Education, Ltd.

To learn more about Brooks/Cole, visit
**www.cengage.com/ brookscole**

Purchase any of our products at your local college store or at our preferred online store
**www.ichapters.com**

Printed in Canada
1 2 3 4 5 6 7 13 12 11 10 09

# Table of Contents

© 2011 Cengage Learning. All Rights Reserved. May not be scanned, copied or duplicated, or posted to a publicly accessible website, in whole or in part.

# Chapter 1    Preparation for Calculus

Course Number

Instructor

Date

## Section 1.1  Graphs and Models

**Objective:** In this lesson you learned how to identify the characteristics of equations and sketch their graphs.

---

**Important Vocabulary**          Define each term or concept.

**Graph of an equation**  _the set of all solution points_

**Intercepts** _points at which the graph intercepts the x or y axis_

---

### I. The Graph of an Equation  (Pages 2–3)

The point (1, 3) is a ___solution point___ of the equation

$-4x + 3y = 5$ because the equation is satisfied when 1 is

substituted for ___x___ and 3 is substituted for ___y___.

**What you should learn**
How to sketch the graph
of an equation

To sketch the graph of an equation using the point-plotting

method, _solve the original equation_
_for y. Then construct a table of_
_values by substituting several_
_variables for x. Plot the points_
_shown in the table_

One disadvantage of the point-plotting method is _that_
_many points may need to be_
_plotted to have a clear idea_
_of the shape of the graph._

**Example 1:**  Complete the table. Then use the resulting solution
points to sketch the graph of the equation
$y = 3 - 0.5x$.

| $x$ | $-4$ | $-2$ | 0 | 2 | 4 |
|---|---|---|---|---|---|
| $y$ | $-6$ | $-3$ | $0$ | $3$ | $6$ |

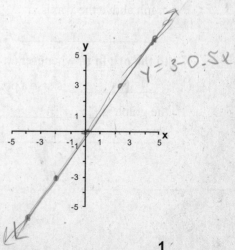

$y = 3 - 0.5x$

---

© 2011 Cengage Learning. All Rights Reserved. May not be scanned, copied or duplicated, or posted to a publicly accessible website, in whole or in part.

## II. Intercepts of a Graph  (Page 4)

The point $(a, 0)$ is a(n) ___*x -intercept*___ of the graph

of an equation if it is a solution point of the equation. The point

$(0, b)$ is a(n) ___*y-intercept*___ of the graph of an

equation if it is a solution point of the equation.

**What you should learn**
How to find the
intercepts of a graph

To find the $x$-intercepts of a graph, ___*let y be 0*___
___*and solve for x*___
To find the $y$-intercepts of a graph, ___*let x be 0*___
___*and solve for y*___

## III. Symmetry of a Graph  (Pages 5–6)

Knowing the symmetry of a graph before attempting to sketch it

is useful because ___*only half as many*___
___*points are needed to sketch*___
___*the graph*___

The three types of symmetry that a graph can exhibit are

___*y-axis, x-axis, origin*___

**What you should learn**
How to test a graph for
symmetry with respect to
an axis and the origin

A graph is **symmetric with respect to the $y$-axis** if, whenever

$(x, y)$ is a point on the graph, ___$(-x, y)$___ is also a point on the

graph. This means that the portion of the graph to the left of the

$y$-axis is ___*a mirror image to the right of*___
___*the y-axis*___. A graph is **symmetric with respect to the**

**$x$-axis** if, whenever $(x, y)$ is a point on the graph, ___$(x, -y)$___ is

also a point on the graph. This means that the portion of the

graph above the $x$-axis is ___*is a mirror image to the*___
___*portion below*___
___*the x-axis*___. A graph is **symmetric with respect**

**to the origin** if, whenever $(x, y)$ is a point on the graph,

___$(-x, -y)$___ is also a point on the graph. This means that

the graph is ___*unchanged by a 180°*___
___*rotation about the origin*___.

© 2011 Cengage Learning. All Rights Reserved. May not be scanned, copied or duplicated, or posted to a publicly accessible website, in whole or in part.

The graph of an equation in $x$ and $y$ is symmetric with respect to the $y$-axis if <u>replacing x w/ -x yields an equivalent equation</u>.

The graph of an equation in $x$ and $y$ is symmetric with respect to the $x$-axis if <u>replacing y w/ -y yields an equivalent equation</u>.

The graph of an equation in $x$ and $y$ is symmetric with respect to the origin if <u>replacing x w/ -x AND y w/ -y yields an equivalent equation</u>.

**Example 2:**   Use symmetry to sketch the graph of the equation $y = 2x^2 + 2$.

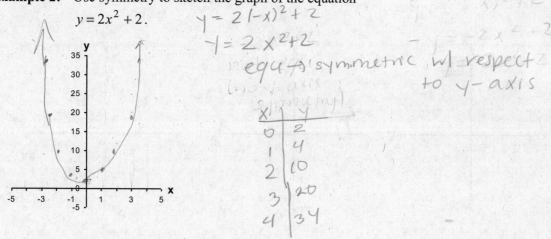

$y = 2(-x)^2 + 2$
$y = 2x^2 + 2$
equ'n symmetric w/ respect to y-axis
(no x-axis symmetry)

| x | y |
|---|---|
| 0 | 2 |
| 1 | 4 |
| 2 | 10 |
| 3 | 20 |
| 4 | 34 |

## IV. Points of Intersection  (Page 6)

<table>
<tr><td>

A **point of intersection** of the graphs of two equations is <u>a point that satisfies both equations</u>

You can find the points of intersection of two graphs by <u>solving their equations simultaneously</u>

</td><td>

**What you should learn**
How to find the points of intersection of two graphs

</td></tr>
</table>

**Example 3:**   Find the point of intersection of the graphs of $y = 2x + 10$ and $y = 14 - 3x$.

$2x + 10 = 14 - 3x$
$5x - 4 = 0$
$5x = 4$
$x = \dfrac{4}{5}$

© 2011 Cengage Learning. All Rights Reserved. May not be scanned, copied or duplicated, or posted to a publicly accessible website, in whole or in part.

## V. Mathematical Models (Page 7)

In developing a mathematical model to represent actual data, strive for two (often conflicting) goals: <u>accuracy</u> <u>and simplicity</u>.

> ***What you should learn***
> How to interpret mathematical models for real-life data

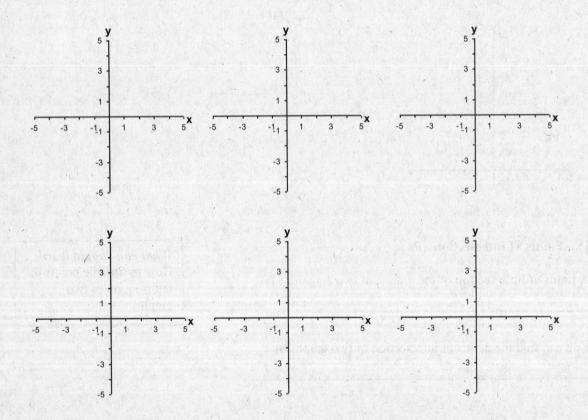

**Homework Assignment**

Page(s)

Exercises

© 2011 Cengage Learning. All Rights Reserved. May not be scanned, copied or duplicated, or posted to a publicly accessible website, in whole or in part.

## Section 1.2  Linear Models and Rates of Change

**Objective:**   In this lesson you learned how to find and graph
equations of lines, including parallel and perpendicular
lines, using the concept of slope.

Course Number

Instructor

Date

**Important Vocabulary**          Define each term or concept.

**Slope**

**Parallel**

**Perpendicular**

### I.  The Slope of a Line  (Page 10)

The **slope** of the nonvertical line passing through the points
$(x_1, y_1)$ and $(x_2, y_2)$ is $m =$ _____.

To find the slope of the line through the points $(-2, 5)$ and
$(4, -3)$, _____

_____

If a line falls from left to right, it has _____ slope. If a
line is horizontal, it has _____ slope. If a line is
vertical, it has _____ slope. If a line rises from left to
right, it has _____ slope.

*What you should learn*
How to find the slope of
a line passing through
two points

### II.  Equations of Lines  (Page 11)

The **point-slope equation of a line** with slope $m$, passing
through the point $(x_1, y_1)$ is

_____ .

*What you should learn*
How to write the
equation of a line given a
point and the slope

**Example 1:**   Find an equation of the line that passes through the
points $(1, 5)$ and $(-3, 7)$.

© 2011 Cengage Learning. All Rights Reserved. May not be scanned, copied or duplicated, or posted to a publicly accessible website, in whole or in part.

### III. Ratios and Rates of Change (Page 12)

In real-life problems, the slope of a line can be interpreted as

either _____, if the x-axis and y-axis have the same

unit of measure, or _____, if the x-axis

and y-axis have different units of measure.

| *What you should learn* |
|---|
| How to interpret slope as a ratio or as a rate in a real-life application |

An **average rate of change** is always calculated over _____

_____.

### IV. Graphing Linear Models (Pages 13–14)

The **slope-intercept** form of the equation of a line is

_____. The graph of this equation is a line having

a slope of _____ and a y-intercept at (____, ____).

| *What you should learn* |
|---|
| How to sketch the graph of a linear equation in slope-intercept form |

**Example 2:**  Explain how to graph the linear equation
$y = -2/3x - 4$. Then sketch its graph.

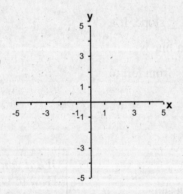

**Example 3:**  Sketch and label the graph of (a) $y = -1$ and
(b) $x = 3$.

(a)                                           (b)

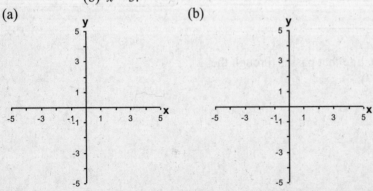

© 2011 Cengage Learning. All Rights Reserved. May not be scanned, copied or duplicated, or posted to a publicly accessible website, in whole or in part.

The equation of a vertical line cannot be written in slope-
intercept form because _____

_____. A vertical

line has an equation of the form _____.

The equation of any line can be written in **general form,** which
is given as _____, where *A* and *B*
are not both zero.

## V.  Parallel and Perpendicular Lines  (Page 14–15)

The relationship between the slopes of two lines that are parallel
is _____

The relationship between the slopes of two lines that are

perpendicular is _____

_____

A line that is parallel to a line whose slope is 2 has slope _____.

A line that is perpendicular to a line whose slope is 2 has slope

_____.

> ***What you should learn***
> How to write equations
> of lines that are parallel
> or perpendicular to a
> given line

© 2011 Cengage Learning. All Rights Reserved. May not be scanned, copied or duplicated, or posted to a publicly accessible website, in whole or in part.

**Additional notes**

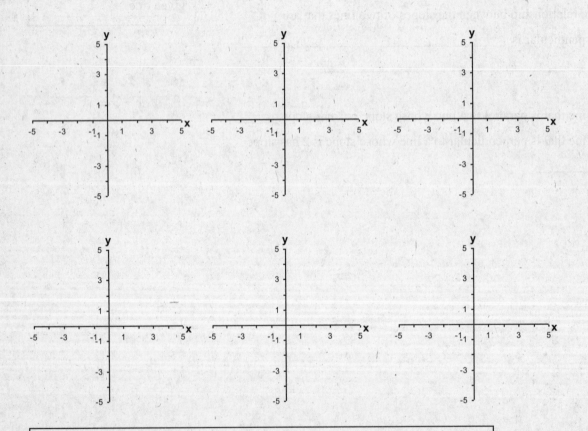

┌─────────────────────────────────────────────────┐
│  **Homework Assignment**                          │
│                                                   │
│  Page(s)                                          │
│                                                   │
│  Exercises                                        │
└─────────────────────────────────────────────────┘

© 2011 Cengage Learning. All Rights Reserved. May not be scanned, copied or duplicated, or posted to a publicly accessible website, in whole or in part.

## Section 1.3  Functions and Their Graphs

**Objective:**      In this lesson you learned how to evaluate and graph
                    functions and their transformations.

Course Number

Instructor

Date

---

**Important Vocabulary**          Define each term or concept.

**Independent variable**

**Dependent variable**

**Function**

---

**I.  Functions and Function Notation**  (Pages 19–20)

Let $X$ and $Y$ be sets of real numbers. A **real-valued function $f$ of
a real variable $x$** from $X$ to $Y$ is _____

_____ .

In this situation, the **domain** of $f$ is _____ . The
number $y$ is the _____ of $x$ under $f$ and is denoted by

_____ , which is called the **value of $f$ at $x$.** The
**range** of $f$ is _____ and consists of

_____ .

In the function  $y = 2 + 8x - 3x^2$ , which variable is the

independent variable? _____
Which variable is the dependent variable? _____

**Example 1:**  If  $f(w) = 4w^3 - 5w^2 - 7w + 13$ , describe how to
              find  $f(-2)$  and then find the value of  $f(-2)$ .

*What you should learn*
How to use function
notation to represent and
evaluate a function

---

Larson/Edwards   **Calculus:  Early Transcendental Functions 5e**   Notetaking Guide

© 2011 Cengage Learning. All Rights Reserved. May not be scanned, copied or duplicated, or posted to a publicly accessible website, in whole or in part.

## II.  The Domain and Range of a Function  (Page 21)

**What you should learn**
How to find the domain
and range of a function

The domain of a function can be described explicitly, or it may

be described implicitly by _____

_____. The implied domain is _____

_____,

whereas an explicitly defined domain is one that is _____

_____.

A function from $X$ to $Y$ is **one-to-one** if _____

_____

_____

A function from $X$ to $Y$ is **onto** if _____

_____

## III.  The Graph of a Function  (Page 22)

**What you should learn**
How to sketch the graph
of a function

The graph of the function $y = f(x)$ consists of _____

_____.

The **Vertical Line Test** states that _____

_____

_____

**Example 2:**   Decide whether each graph represents $y$ as a
                    function of $x$.

**(a)**                                          **(b)**

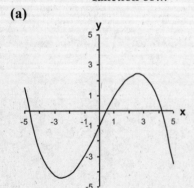                              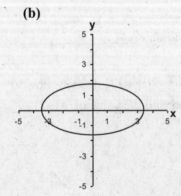

© 2011 Cengage Learning. All Rights Reserved. May not be scanned, copied or duplicated, or posted to a publicly accessible website, in whole or in part.

Sketch an example of each of the following eight basic graphs.

Squaring Function

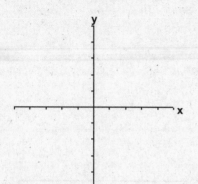

Identity Function

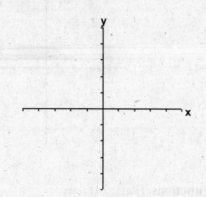

Absolute Value Function

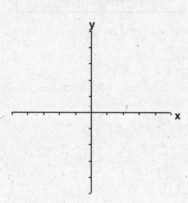

Square Root Function

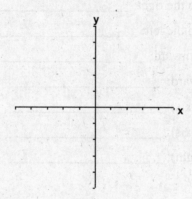

Rational Function

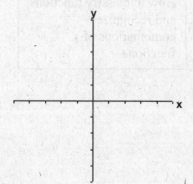

Cubing Function

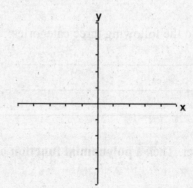

© 2011 Cengage Learning. All Rights Reserved. May not be scanned, copied or duplicated, or posted to a publicly accessible website, in whole or in part.

Sine Function                              Cosine Function

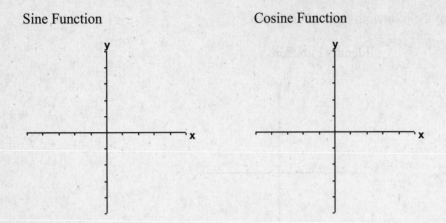

## IV.  Transformations of Functions (Page 23)

Let $c$ be a positive real number. Complete the following
representations of shifts in the graph of $y = f(x)$:

1)  Horizontal shift $c$ units to the right: _____

2)  Horizontal shift $c$ units to the left: _____

3)  Vertical shift $c$ units downward: _____

4)  Vertical shift $c$ units upward: _____

5)  Reflection (about the $x$-axis): _____

6)  Reflection (about the $y$-axis): _____

7)  Reflection (about the origin): _____

> **What you should learn**
> How to identify different
> types of transformations
> of functions

## V.  Classifications and Combinations of Functions
   (Pages 24–26)

Elementary functions fall into the following three categories:

_____

_____

_____.

> **What you should learn**
> How to classify functions
> and recognize
> combinations of
> functions

Let $n$ be a nonnegative integer. Then a **polynomial function of $x$**
is given as

_____

The numbers $a_i$ are _____, with $a_n$ the _____

_____ and $a_0$ the _____ of the

polynomial function. If $a_n \neq 0$, then $n$ is the _____ of

the polynomial function.

© 2011 Cengage Learning. All Rights Reserved. May not be scanned, copied or duplicated, or posted to a publicly accessible website, in whole or in part.

Just as a rational number can be written as the quotient of two
integers, a rational function can be written as _____
_____.

An algebraic function of $x$ is one that _____
_____
_____. Functions that are
not algebraic are _____.

Two functions can be combined by the operations of
_____
to create new functions.

Functions can also be combined through **composition.** The
resulting function is called a(n) _____.

Let $f$ and $g$ be functions. The function given by $(f \circ g)(x) =$
_____ is called the **composite** of $f$ with $g$. The
domain of $f \circ g$ is _____
_____.

**Example 3:**    Let $f(x) = 3x + 4$ and let $g(x) = 2x^2 - 1$. Find
(a) $(f \circ g)(x)$ and (b) $(g \circ f)(x)$.

An $x$-intercept of a graph is defined to be a point $(a, 0)$ at which
the graph crosses the $x$-axis. If the graph represents a function $f$,
the number $a$ is a _____. In other words,
the zeros of a function $f$ are _____
_____.

A function is **even** if _____
_____. A function is **odd** if
_____.

© 2011 Cengage Learning. All Rights Reserved. May not be scanned, copied or duplicated, or posted to a publicly accessible website, in whole or in part.

The function $y = f(x)$ is **even** if _____.

The function $y = f(x)$ is **odd** if _____.

**Example 4:** Decide whether the function $f(x) = 4x^2 - 3x + 1$ is even, odd, or neither.

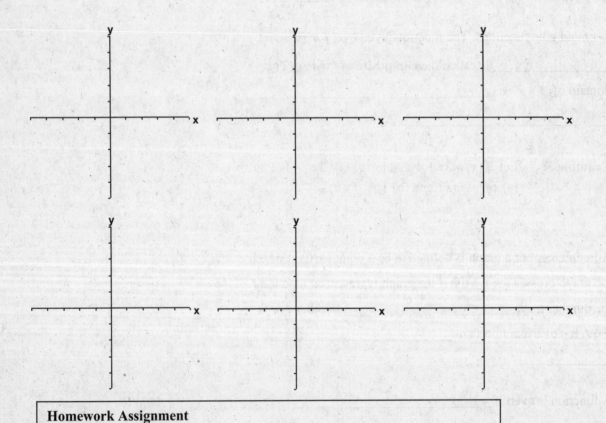

**Homework Assignment**

Page(s)

Exercises

© 2011 Cengage Learning. All Rights Reserved. May not be scanned, copied or duplicated, or posted to a publicly accessible website, in whole or in part.

## Section 1.4  Fitting Models to Data

**Objective:**   In this lesson you learned how to fit mathematical
models to real-life data sets.

Course Number

Instructor

Date

**I. Fitting a Linear Model to Data**  (Page 31)

Describe how to find a linear model to represent a set of paired
data.

***What you should learn***
How to fit a linear model
to a real-life data set

What does the correlation coefficient $r$ indicate?

**Example 1:**  Find a linear model to represent the following
data. Round results to the nearest hundredth.

| | | |
|---|---|---|
| (−2.1, 19.4) | (−3.0, 19.7) | (8.8, 16.9) |
| (0, 18.9) | (6.1, 17.4) | (−4.0, 20.0) |
| (3.6, 18.1) | (0.9, 18.8) | (2.0, 18.5) |

© 2011 Cengage Learning. All Rights Reserved. May not be scanned, copied or duplicated, or posted to a publicly accessible website, in whole or in part.

## II. Fitting a Quadratic Model to Data (Page 32)

**Example 2:** Find a model to represent the following data. Round results to the nearest hundredth.

| | | |
|---|---|---|
| (−5, 68) | (−3, 30) | (−2, 22) |
| (−1, 11) | (0, 3) | (2, 8) |
| (4, 23) | (5, 43) | (7, 80) |

> ***What you should learn***
> How to fit a quadratic model to a real-life data set

## III. Fitting a Trigonometric Model to Data (Page 33)

**Example 3:** Find a trigonometric function to model the data in the following table.

| $x$ | 0 | $\pi/2$ | $\pi$ | $3\pi/2$ | $2\pi$ |
|---|---|---|---|---|---|
| $y$ | 2 | 4 | 2 | 0 | 2 |

> ***What you should learn***
> How to fit a trigonometric model to a real-life data set

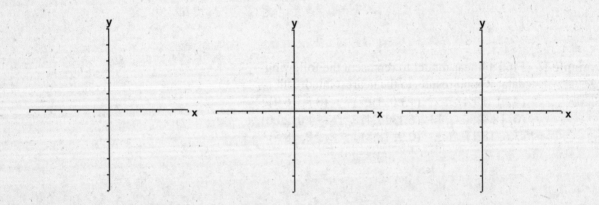

---

**Homework Assignment**

Page(s)

Exercises

---

© 2011 Cengage Learning. All Rights Reserved. May not be scanned, copied or duplicated, or posted to a publicly accessible website, in whole or in part.

## Section 1.5  Inverse Functions

**Objective:**     In this lesson you learned how to determine whether a function has an inverse function and how to identify the properties of inverse trigonometric functions.

Course Number

Instructor

Date

---

| **Important Vocabulary** | Define each term or concept. |
| --- | --- |

**Inverse function**


**Horizontal Line Test**

---

### I. Inverse Functions  (Pages 37–38)

For a function $f$ that is represented by a set of ordered pairs, you can form the inverse function of $f$ by _____

_____ .

*What you should learn*
How to verify that one function is the inverse function of another function

For a function $f$ and its inverse $f^{-1}$, the domain of $f$ is equal to

_____ , and the range of $f$ is equal to

_____ .

State three important observations about inverse functions.

1.



2.



3.



To verify that two functions, $f$ and $g$, are inverse functions of each other, . . .

© 2011 Cengage Learning. All Rights Reserved. May not be scanned, copied or duplicated, or posted to a publicly accessible website, in whole or in part.

**Example 1:**    Verify that the functions $f(x) = 2x - 3$ and

$g(x) = \dfrac{x+3}{2}$ are inverse functions of each other.


The graph of $f^{-1}$ is a **reflection** of the graph of $f$ in the line

_____.


The Reflective Property of Inverse Functions states that the

graph of $f$ contains the point $(a, b)$ if and only if _____

_____.


## II. Existence of an Inverse Function  (Pages 39–40)

Explain why the horizontal line test is valid.

| *What you should learn* |
| :--- |
| How to determine whether a function has an inverse function |


**Example 2:**    Does the graph of the function shown below
have an inverse function? Explain.

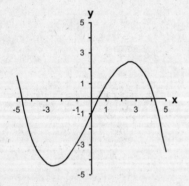

Complete the following guidelines for finding an inverse

function.

1)


2)

3)

© 2011 Cengage Learning. All Rights Reserved. May not be scanned, copied or duplicated, or posted to a publicly accessible website, in whole or in part.

4)

5)

**Example 3:**    Find the inverse (if it exists) of $f(x) = 4x - 5$.

### III. Inverse Trigonometric Functions  (Pages 41–43)

None of the six basic trigonometric functions has _____

_____. This is true because all six

trigonometric functions are _____

_____. The functions that are called

"inverse trigonometric functions" are actually _____

_____.

*What you should learn*
How to develop
properties of the six
inverse trigonometric
functions

For each of the following definitions of inverse trigonometric
functions, supply the required restricted domains and ranges.

|  | Domain | Range |
|---|---|---|
| $y = \arcsin x$ iff $\sin y = x$ | _____ | _____ |
| $y = \arccos x$ iff $\cos y = x$ | _____ | _____ |
| $y = \arctan x$ iff $\tan y = x$ | _____ | _____ |
| $y = \text{arccot } x$ iff $\cot y = x$ | _____ | _____ |
| $y = \text{arcsec } x$ iff $\sec y = x$ | _____ | _____ |
| $y = \text{arccsc } x$ iff $\csc y = x$ | _____ | _____ |

An alternative notation for the inverse sine function is

_____.

**Example 4:**    Evaluate the function:  $\arcsin(-1)$.

© 2011 Cengage Learning. All Rights Reserved. May not be scanned, copied or duplicated, or posted to a publicly accessible website, in whole or in part.

**Example 5:**   Evaluate the function:  $\arccos \dfrac{1}{2}$.

**Example 6:**   Evaluate the function:  arcos (0.85).

State the Inverse Property for the Sine function.

State the Inverse Property for the Tangent function.

State the Inverse Property for the Secant function.

```
Homework Assignment

Page(s)

Exercises
```

© 2011 Cengage Learning. All Rights Reserved. May not be scanned, copied or duplicated, or posted to a publicly accessible website, in whole or in part.

## Section 1.6   Exponential and Logarithmic Functions

**Objective:**   In this lesson you learned about the properties of the natural exponential and natural logarithmic functions.

| Course Number |
| Instructor |
| Date |

**I. Exponential Functions** (Pages 49–50)

An **exponential function** involves _____

_____.

The exponential function with base $a$ is written as _____

_____.

*What you should learn*
How to develop and use
properties of exponential
functions

Let $a$ and $b$ be positive real numbers, and let $x$ and $y$ be any real numbers. Complete the following properties of exponents.

$(ab)^x =$ _____          $a^{x+y} =$ _____

$\dfrac{1}{a^x} =$ _____          $\left(\dfrac{a}{b}\right)^x =$ _____

$\dfrac{a^x}{a^y} =$ _____          $a^{xy} =$ _____

$a^0 =$ _____

**Example 1:**   Sketch the graph of the function $f(x) = 3^{-x}$.

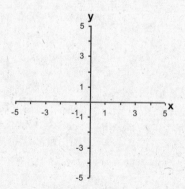

For $a > 1$, is the graph of $f(x) = a^x$ increasing or decreasing over its domain? _____

For $a > 1$, is the graph of $g(x) = a^{-x}$ increasing or decreasing over its domain? _____

© 2011 Cengage Learning. All Rights Reserved. May not be scanned, copied or duplicated, or posted to a publicly accessible website, in whole or in part.

For the graph of $f(x) = a^x$ or $g(x) = a^{-x}$, $a > 1$, the domain is

_____, the range is _____, and

the $y$-intercept is _____. Also, both graphs are

_____.

## II. The Number $e$ (Page 51)

The number $e$ is a(n) _____ number,

which has the decimal approximation _____.

**What you should learn**
How to understand the definition of the number $e$

## III. The Natural Logarithmic Function (Pages 51–52)

Because the natural exponential function $f(x) = e^x$ is one-to-one,

it must have a(n) _____. This is

called the _____.

**What you should learn**
How to understand the definition of the natural logarithmic function

Let $x$ be a positive real number. The **natural logarithmic
function,** denoted by $\ln x$, is defined as $\ln x = b$ if and only if

_____.

Because the function $g(x) = \ln x$ is defined to be the inverse of

$f(x) = e^x$, it follows that the graph of the natural logarithmic

function is _____

_____.

Is the graph of $g(x) = \ln x$ increasing or decreasing over its

domain? _____

For the function $g(x) = \ln x$, the domain is _____,

the range is _____, and the $x$-intercept is

_____. Furthermore, the function $g(x) = \ln x$ is

_____.

© 2011 Cengage Learning. All Rights Reserved. May not be scanned, copied or duplicated, or posted to a publicly accessible website, in whole or in part.

Because $f(x) = e^x$ and $g(x) = \ln x$ are inverses of each other, you

can conclude that $\ln e^x =$ _____ and $e^{\ln x} =$ _____.


### IV. Properties of Logarithms (Pages 53–54)

Let $x$, $y$, and $z$ be real numbers such that $x > 0$ and $y > 0$, then the
following properties are true:

1. $\ln xy =$ _____.

2. $\ln x^z =$ _____.

3. $\ln\left(\dfrac{x}{y}\right) =$ _____.

**What you should learn**
How to develop and use
properties of the natural
logarithmic function

**Example 2:** Expand the logarithmic expression $\ln \dfrac{xy^4}{2}$.




When using the properties of logarithms to rewrite logarithmic

functions, you must _____

_____

_____.




**Example 3:** Solve $e^{x-2} - 7 = 59$ for $x$. Round to three decimal
places.

© 2011 Cengage Learning. All Rights Reserved. May not be scanned, copied or duplicated, or posted to a publicly accessible website, in whole or in part.

**Example 4:**   Solve $4 \ln 5x = 28$ for $x$. Round to three decimal
places.

**Additional notes**

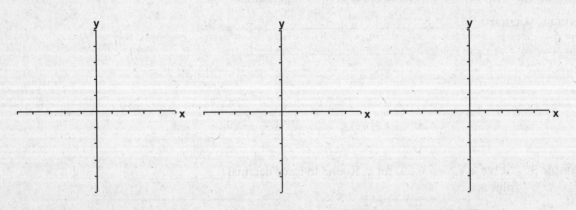

---

**Homework Assignment**

Page(s)

Exercises

© 2011 Cengage Learning. All Rights Reserved. May not be scanned, copied or duplicated, or posted to a publicly accessible website, in whole or in part.

# Chapter 2    Limits and Their Properties

Course Number

Instructor

Date

## Section 2.1   A Preview of Calculus

**Objective:** In this lesson you learned how calculus compares with precalculus.

### I.  What is Calculus?  (Pages 62–64)

Calculus is _____

_____

_____

_____

_____ .

List some problem-solving strategies that will be helpful in the study of calculus.

> **What you should learn**
> How to understand what calculus is and how it compares with precalculus

### II.  The Tangent Line Problem  (Page 65)

In the tangent line problem, you are given _____

_____ and are asked to _____

_____ .

Except for cases involving a vertical tangent line, the problem of

finding the tangent line at a point $P$ is equivalent to _____

_____ . You can

approximate this slope by using a line through _____

_____ . Such a

line is called a _____ .

> **What you should learn**
> How to understand that the tangent line problem is basic to calculus

© 2011 Cengage Learning. All Rights Reserved. May not be scanned, copied or duplicated, or posted to a publicly accessible website, in whole or in part.

If $P(c, f(c))$ is the point of tangency and $Q(c + \Delta x, f(c + \Delta x))$ is a

second point on the graph of $f$, the slope of the secant line

through these two points can be found using precalculus and is

given by $m_{sec} = \underline{\hspace{5cm}}$.

As point $Q$ approaches point $P$, the slope of the secant line

approaches the slope of the $\underline{\hspace{5cm}}$.

When such a "limiting position" exists, the slope of the tangent

line is said to be $\underline{\hspace{8cm}}$

$\underline{\hspace{4cm}}$.

### III. The Area Problem (Page 66)

A second classic problem in calculus is $\underline{\hspace{4cm}}$

$\underline{\hspace{9cm}}$

$\underline{\hspace{4cm}}$. This problem can also be solved

with $\underline{\hspace{5cm}}$. In this case, the limit

process is applied to $\underline{\hspace{5cm}}$

$\underline{\hspace{5cm}}$.

<table><tr><td>**What you should learn**<br>How to understand that the area problem is also basic to calculus</td></tr></table>

Consider the region bounded by the graph of the function

$y = f(x)$, the $x$-axis, and the vertical lines $x = a$ and $x = b$. You

can approximate the area of the region with $\underline{\hspace{3cm}}$

$\underline{\hspace{5cm}}$. As you increase the number

of rectangles, the approximation tends to become $\underline{\hspace{2cm}}$

$\underline{\hspace{6cm}}$

$\underline{\hspace{4cm}}$. Your goal is to determine the

limit of the sum of the areas of the rectangles as $\underline{\hspace{3cm}}$

$\underline{\hspace{6cm}}$.

**Homework Assignment**

Page(s)

Exercises

© 2011 Cengage Learning. All Rights Reserved. May not be scanned, copied or duplicated, or posted to a publicly accessible website, in whole or in part.

## Section 2.2  Finding Limits Graphically and Numerically

**Objective:**      In this lesson you learned how to find limits graphically and numerically.

Course Number

Instructor

Date

### I.  An Introduction to Limits  (Pages 68–69)

The notation for a limit is $\lim\limits_{x \to c} f(x) = L$, which is read as

_____

The informal description of a limit is as follows: _____

_____

_____.

Describe how to estimate the limit $\lim\limits_{x \to -2} \dfrac{x^2 + 4x + 4}{x + 2}$ numerically.

The existence or nonexistence of $f(x)$ at $x = c$ has no bearing on the existence of _____.

*What you should learn*
How to estimate a limit using a numerical or graphical approach

### II.  Limits That Fail to Exist  (Pages 70–71)

If a function $f(x)$ approaches a different number from the right side of $x = c$ than it approaches from the left side, then _____

_____.

If $f(x)$ is not approaching a real number $L$—that is, if $f(x)$ increases or decreases without bound—as $x$ approaches $c$, you can conclude that _____.

The limit of $f(x)$ as $x$ approaches $c$ also does not exist if $f(x)$ oscillates between _____ as $x$ approaches $c$.

*What you should learn*
How to learn different ways that a limit can fail to exist

© 2011 Cengage Learning. All Rights Reserved. May not be scanned, copied or duplicated, or posted to a publicly accessible website, in whole or in part.

### III.  A Formal Definition of Limit  (Pages 72–74)

| *What you should learn* |
| --- |
| How to study and use a formal definition of limit |

The **ε-δ definition of limit** assigns mathematically rigorous

meanings to the two phrases _____

_____ and _____ used

in the informal description of limit.

Let ε represent _____. Then the

phrase "$f(x)$ becomes arbitrarily close to $L$" means that $f(x)$ lies in

the interval _____. Using absolute value,

you can write this as _____. The phrase

"$x$ approaches $c$" means that there exists a positive number δ

such that $x$ lies in either the interval _____ or the

interval _____. This fact can be concisely

expressed by the double inequality _____.

State the formal ε-δ definition of limit.

**Example 1:**   Use the ε-δ definition of limit to prove that
$$\lim_{x \to -2}(10 - 3x) = 16.$$

| **Homework Assignment** |
| --- |
| Page(s) |
| Exercises |

© 2011 Cengage Learning. All Rights Reserved. May not be scanned, copied or duplicated, or posted to a publicly accessible website, in whole or in part.

## Section 2.3  Evaluating Limits Analytically

**Objective:**    In this lesson you learned how to evaluate limits
analytically.

Course Number

Instructor

Date

### I.  Properties of Limits  (Pages 79–81)

The limit of $f(x)$ as $x$ approaches $c$ does not depend on the value
of $f$ at $x = c$. However, it may happen that the limit is precisely
$f(c)$. In such cases, the limit can be evaluated by _____
_____.

*What you should learn*
How to evaluate a limit
using properties of limits

**Theorem 2.1**  Let $b$ and $c$ be real numbers and let $n$ be a positive
integer. Complete each of the following properties of limits.

1.  $\lim\limits_{x \to c} b =$ _____

2.  $\lim\limits_{x \to c} x =$ _____

3.  $\lim\limits_{x \to c} x^n =$ _____

**Theorem 2.2**  Let $b$ and $c$ be real numbers, let $n$ be a positive
integer, and let $f$ and $g$ be functions with the following limits.

$$\lim_{x \to c} f(x) = L \quad \text{and} \quad \lim_{x \to c} g(x) = K$$

Complete each of the following statements about operations with
limits.

1.  Scalar multiple:    $\lim\limits_{x \to c} [b\, f(x)] =$ _____

2.  Sum or difference:    $\lim\limits_{x \to c} [f(x) \pm g(x)] =$ _____

3.  Product:    $\lim\limits_{x \to c} [f(x) \cdot g(x)] =$ _____

4.  Quotient:    $\lim\limits_{x \to c} \dfrac{f(x)}{g(x)} =$ _____

5.  Power:    $\lim\limits_{x \to c} [f(x)]^n =$ _____

**Example 1:**  Find the limit: $\lim\limits_{x \to 4} 3x^2$.

Larson/Edwards  **Calculus: Early Transcendental Functions 5e**  Notetaking Guide

© 2011 Cengage Learning. All Rights Reserved. May not be scanned, copied or duplicated, or posted to a publicly accessible website, in whole or in part.

The limit of a polynomial function $p(x)$ as $x \to c$ is simply the value of $p$ at $x = c$. This direction substitution property is valid

for _____

_____ .

**Theorem 2.3**  If $p$ is a polynomial function and $c$ is a real

number, then $\lim_{x \to c} p(x) =$ _____ . If $r$ is a rational

function given by $r(x) = p(x)/q(x)$ and $c$ is a real number such

that $q(c) \neq 0$, then $\lim_{x \to c} r(x) =$ _____ .

**Theorem 2.4**  Let $n$ be a positive integer. The following limit is valid for all $c$ if $n$ is odd, and is valid for $c > 0$ if $n$ is even:

$$\lim_{x \to c} \sqrt[n]{x} = \text{_____}$$

**Theorem 2.5**  If $f$ and $g$ are functions such that $\lim_{x \to c} g(x) = L$ and

$\lim_{x \to L} f(x) = f(L)$, then $\lim_{x \to c} f(g(x)) =$ _____ .

**Theorem 2.6**  Let $c$ be a real number in the domain of the given transcendental function. Complete each of the following limit statements.

1. $\lim_{x \to c} \sin x =$ _____

2. $\lim_{x \to c} \cos x =$ _____

3. $\lim_{x \to c} \tan x =$ _____

4. $\lim_{x \to c} \cot x =$ _____

5. $\lim_{x \to c} \sec x =$ _____

6. $\lim_{x \to c} \csc x =$ _____

7. $\lim_{x \to c} a^x =$ _____

8. $\lim_{x \to c} \ln x =$ _____

© 2011 Cengage Learning. All Rights Reserved. May not be scanned, copied or duplicated, or posted to a publicly accessible website, in whole or in part.

**Example 2:**    Find the following limits.

a.  $\lim\limits_{x \to 4} \sqrt[4]{5x^2 + 1}$          b.  $\lim\limits_{x \to \pi} \cos x$          c.  $\lim\limits_{x \to 2} \ln x$

## II.  A Strategy for Finding Limits  (Page 82)

> *What you should learn*
> How to develop and use a
> strategy for finding limits

**Theorem 2.7**  Let $c$ be a real number and let $f(x) = g(x)$ for all

$x \neq c$ in an open interval containing $c$. If the limit of $g(x)$ as $x$

approaches $c$ exists, then the limit of $f(x)$ _____ and

$\lim\limits_{x \to c} f(x) =$ _____.

List four steps in the strategy for finding limits.

## III.  Dividing Out and Rationalizing Techniques
   (Pages 83–84)

> *What you should learn*
> How to evaluate a limit
> using dividing out and
> rationalizing techniques

An expression such as the meaningless fractional form 0/0 is

called a(n) _____ because you

cannot, from the form alone, determine the limit. When you try

to evaluate a limit and encounter this form, remember that you

must rewrite the fraction so that the new denominator _____

_____. One way to do this is to _____

_____, using the **dividing out**

technique. Another technique is to _____ the

numerator.

**Example 3:**    Find the following limit:  $\lim\limits_{x \to 3} \dfrac{x^2 - 8x + 15}{x - 3}$.

Larson/Edwards  **Calculus: Early Transcendental Functions 5e**  Notetaking Guide

© 2011 Cengage Learning. All Rights Reserved. May not be scanned, copied or duplicated, or posted to a publicly accessible website, in whole or in part.

If you apply direct substitution to a rational function and obtain

$r(c) = \dfrac{p(c)}{q(c)} = \dfrac{0}{0}$, then by the Factor Theorem of Algebra, you

can conclude that $(x - c)$ must be a _____ to

both $p(x)$ and $q(x)$.

**IV. The Squeeze Theorem**  (Pages 85–86)

**Theorem 2.8  The Squeeze Theorem**  If $h(x) \le f(x) \le g(x)$  for

all $x$ in an open interval containing $c$, except possibly at $c$ itself,

and if $\lim\limits_{x \to c} h(x) = L = \lim\limits_{x \to c} g(x)$, then $\lim\limits_{x \to c} f(x)$  exists and is

equal to _____.

> **What you should learn**
> How to evaluate a limit using the Squeeze Theorem

**Theorem 2.9  Three Special Limits**

$$\lim_{x \to 0} \frac{\sin x}{x} = \underline{\hspace{2cm}}$$

$$\lim_{x \to 0} \frac{1 - \cos x}{x} = \underline{\hspace{2cm}}$$

$$\lim_{x \to 0} (1 + x)^{1/x} = \underline{\hspace{2cm}}$$

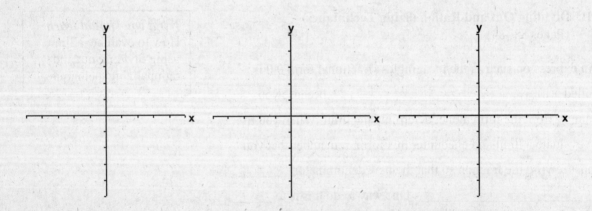

**Homework Assignment**

Page(s)

Exercises

© 2011 Cengage Learning. All Rights Reserved. May not be scanned, copied or duplicated, or posted to a publicly accessible website, in whole or in part.

## Section 2.4  Continuity and One-Sided Limits

Course Number

Instructor

Date

**Objective:**     In this lesson you learned how to determine continuity at a point and on an open interval, and how to determine one-sided limits.

---

**Important Vocabulary**          Define each term or concept.

**Discontinuity**


**Greatest integer function** $f(x) = [\![x]\!]$

---

### I.  Continuity at a Point and on an Open Interval
(Pages 90–91)

To say that a function $f$ is continuous at $x = c$ means that there is

no _____ in the graph of $f$ at $c$:  the graph is

unbroken and there are no _____.

*What you should learn*
How to determine continuity at a point and continuity on an open interval

A function $f$ is **continuous at $c$** if the following three conditions are met:

1.

2.

3.


If $f$ is continuous at each point in the interval $(a, b)$, then it is

_____. A

function that is continuous on the entire real line $(-\infty, \infty)$ is

_____.


A discontinuity at $c$ is called **removable** if _____

_____

_____.

© 2011 Cengage Learning. All Rights Reserved. May not be scanned, copied or duplicated, or posted to a publicly accessible website, in whole or in part.

A discontinuity at $c$ is called **nonremovable** if _____

_____

_____.


## II. One-Sided Limits and Continuity on a Closed Interval
   (Pages 92–94)

**What you should learn**
How to determine one-
sided limits and
continuity on a closed
interval

A **one-sided limit** is the limit of a function $f(x)$ at $c$ from either

just the _____ of $c$ or just the _____ of $c$.

$\lim\limits_{x \to c^+} f(x) = L$ is a one-sided limit from the _____ and means

_____

$\lim\limits_{x \to c^-} f(x) = L$ is a one-sided limit from the _____ and means

_____

One-sided limits are useful in taking limits of functions

involving _____.

When the limit from the left is not equal to the limit from the

right, the (two-sided) limit _____.

Let $f$ be defined on a closed interval $[a, b]$. If $f$ is continuous on the

open interval $(a, b)$ and $\lim\limits_{x \to a^+} f(x) = f(a)$ and $\lim\limits_{x \to b^-} f(x) = f(b)$,

then $f$ is _____.

Moreover, $f$ is continuous _____ at $a$ and continuous

_____ at $b$.


## III. Properties of Continuity  (Pages 95–96)

**What you should learn**
How to use properties of
continuity

If $b$ is a real number and $f$ and $g$ are continuous at $x = c$, then the
following functions are also continuous at $c$.

1.

2.

3.

4.

© 2011 Cengage Learning. All Rights Reserved. May not be scanned, copied or duplicated, or posted to a publicly accessible website, in whole or in part.

A polynomial function is continuous at _____

_____.

A rational function is continuous at _____

_____.

Radical functions, trigonometric functions, exponential

functions, and logarithmic functions are continuous at _____

_____.

If $g$ is continuous at $c$ and $f$ is continuous at $g(c)$, then the

composite function given by $(f \circ g)(x) = f(g(x))$ is continuous

_____.

## IV.  The Intermediate Value Theorem  (Pages 97–98)

**Intermediate Value Theorem**  If $f$ is continuous on the closed

interval $[a, b]$, $f(a) \neq f(b)$, and $k$ is any number between $f(a)$ and

$f(b)$, then _____

_____.

Explain why the Intermediate Value Theorem is called an

**existence theorem.**

| *What you should learn* |
| How to understand and use the Intermediate Value Theorem |

The Intermediate Value Theorem states that for a continuous

function $f$, if $x$ takes on all values between $a$ and $b$, $f(x)$

must _____

_____.

The Intermediate Value Theorem often can be used to locate the

zeros of a function that is continuous on a closed interval.

© 2011 Cengage Learning. All Rights Reserved. May not be scanned, copied or duplicated, or posted to a publicly accessible website, in whole or in part.

Specifically, if $f$ is continuous on $[a, b]$ and $f(a)$ and $f(b)$ differ in sign, the Intermediate Value Theorem guarantees _____

_____.

Explain how the **bisection method** can be used to approximate the real zeros of a continuous function.

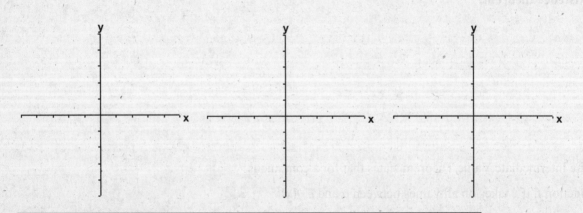

**Homework Assignment**

Page(s)

Exercises

© 2011 Cengage Learning. All Rights Reserved. May not be scanned, copied or duplicated, or posted to a publicly accessible website, in whole or in part.

## Section 2.5  Infinite Limits

**Objective:**    In this lesson you learned how to determine infinite limits and find vertical asymptotes.

Course Number

Instructor

Date

### I. Infinite Limits  (Pages 103–104)

A limit in which $f(x)$ increases or decreases without bound as $x$ approaches $c$ is called a(n) _____.

**What you should learn**
How to determine infinite limits from the left and from the right

Let $f$ be a function that is defined at every real number in some open interval containing $c$ (except possibly at $c$ itself). The statement $\lim_{x \to c} f(x) = \infty$ means _____

_____

_____. Similarly, the statement

$\lim_{x \to c} f(x) = -\infty$ means _____

_____.

To define the **infinite limit from the left,** replace $0 < |x - c| < \delta$ by _____. To define the **infinite limit from the right,** replace $0 < |x - c| < \delta$ by _____.

Be sure to see that the equal sign in the statement $\lim f(x) = \infty$ does not mean that _____! On the contrary, it tells you how the limit _____ by denoting the unbounded behavior of $f(x)$ as $x$ approaches $c$.

### II. Vertical Asymptotes  (Pages 104–107)

If $f(x)$ approaches infinity (or negative infinity) as $x$ approaches $c$ from the right or the left, then the line $x = c$ is a _____ _____ of the graph of $f$.

**What you should learn**
How to find and sketch the vertical asymptotes of the graph of a function

Let $f$ and $g$ be continuous on an open interval containing $c$. If $f(c) \neq 0$, $g(c) = 0$, and there exists an open interval containing $c$ such that $g(x) \neq 0$ for all $x \neq c$ in the interval, then the graph

© 2011 Cengage Learning. All Rights Reserved. May not be scanned, copied or duplicated, or posted to a publicly accessible website, in whole or in part.

of the function given by $h(x) = \dfrac{f(x)}{g(x)}$ has _____

_____ .

If both the numerator and denominator are 0 at $x = c$, you obtain

the _____, and you

cannot determine the limit behavior at $x = c$ without further

investigation, such as simplifying the expression.

**Example 1:**   Determine all vertical asymptotes of the graph of

$$f(x) = \frac{x^2 + 9x + 20}{x^2 + 2x - 15}.$$

**Theorem 2.15**  Let $c$ and $L$ be real numbers and let $f$ and $g$ be
functions such that

$$\lim_{x \to c} f(x) = \infty \quad \text{and} \quad \lim_{x \to c} g(x) = L$$

Complete each of the following statements about operations with
limits.

1. Sum or difference:   $\lim_{x \to c} [f(x) \pm g(x)] = $ _____

2. Product:   $\lim_{x \to c} [f(x) \cdot g(x)] = $ _____

   $\lim_{x \to c} [f(x) \cdot g(x)] = $ _____

3. Quotient:   $\lim_{x \to c} \dfrac{g(x)}{f(x)} = $ _____

**Example 2:**   Determine the limit:   $\lim_{x \to 3} \left( \dfrac{1}{x-3} - 3 \right)$.

---

**Homework Assignment**

Page(s)

Exercises

---

© 2011 Cengage Learning. All Rights Reserved. May not be scanned, copied or duplicated, or posted to a publicly accessible website, in whole or in part.

# Chapter 3     Differentiation

## Section 3.1   The Derivative and the Tangent Line Problem

Course Number

Instructor

Date

**Objective:** In this lesson you learned how to find the derivative of a function using the limit definition and understand the relationship between differentiability and continuity.

---

**Important Vocabulary**      Define each term or concept.

**Differentiation**

**Differentiable**

---

## I. The Tangent Line Problem (Pages 116–119)

**What you should learn**
How to find the slope of the tangent line to a curve at a point

Essentially, the problem of finding the tangent line at a point $P$ boils down to _____

_____. You can approximate this

slope using _____ through the point

of tangency $(c, f(c))$ and a second point on the curve

$(c + \Delta x, f(c + \Delta x))$. The slope of the secant line through these two

points is $m_{sec} = \underline{\hspace{2cm}}$ .

The right side of this equation for the slope of a secant line is

called a _____. The denominator $\Delta x$ is

the _____, and the numerator

$\Delta y = f(c + \Delta x) - f(c)$ is the _____ .

The beauty of this procedure is that you can obtain more and

more accurate approximations of the slope of the tangent line

by _____

_____.

If $f$ is defined on an open interval containing $c$, and if the limit

$$\lim_{\Delta x \to 0} \frac{\Delta y}{\Delta x} = \lim_{\Delta x \to 0} \frac{f(c + \Delta x) - f(c)}{\Delta x} = m \text{ exists, then the line passing}$$

© 2011 Cengage Learning. All Rights Reserved. May not be scanned, copied or duplicated, or posted to a publicly accessible website, in whole or in part.

through $(c, f(c))$ with slope $m$ is _____

_____.

The slope of the tangent line to the graph of $f$ at the point $(c, f(c))$

is also called _____

_____.

**Example 1:**   Find the slope of the graph of $f(x) = 9 - \dfrac{x}{2}$ at the

point (4, 7).

**Example 2:**   Find the slope of the graph of $f(x) = 2 - 3x^2$ at

the point $(-1, -1)$.

The definition of a tangent line to a curve does not cover the

possibility of a vertical tangent line. If $f$ is continuous at $c$ and

$$\lim_{\Delta x \to 0} \frac{f(c + \Delta x) - f(c)}{\Delta x} = \infty \text{ or } \lim_{\Delta x \to 0} \frac{f(c + \Delta x) - f(c)}{\Delta x} = -\infty \text{, the}$$

vertical line $x = c$ passing through $(c, f(c))$ is _____

_____ to the graph of $f$.

**II. The Derivative of a Function**  (Pages 119–121)

The _____ is given by

$$f'(x) = \lim_{\Delta x \to 0} \frac{f(x + \Delta x) - f(x)}{\Delta x} \text{, provided the limit exists. For all } x$$

for which this limit exists, $f'$ is _____.

The derivative of a function of $x$ gives the _____

_____ to the graph of $f$ at the point $(x, f(x))$,

provided that the graph has a tangent line at this point.

A function is **differentiable on an open interval (a, b)** if _____

_____.

> **What you should learn**
> How to use the limit
> definition to find the
> derivative of a function

© 2011 Cengage Learning. All Rights Reserved. May not be scanned, copied or duplicated, or posted to a publicly accessible website, in whole or in part.

**Example 3:**   Find the derivative of $f(t) = 4t^2 + 5$.

### III. Differentiability and Continuity  (Pages 121–123)

Name some situations in which a function will not be
differentiable at a point.

**What you should learn**
How to understand the
relationship between
differentiability and
continuity

If a function $f$ is differentiable at $x = c$, then _____
_____.

Complete the following statements.

1.  If a function is differentiable at $x = c$, then it is continuous at
    $x = c$. So, differentiability _____ continuity.

2.  It is possible for a function to be continuous at $x = c$ and not
    be differentiable at $x = c$. So, continuity _____
    _____ differentiability.

© 2011 Cengage Learning. All Rights Reserved. May not be scanned, copied or duplicated, or posted to a publicly accessible website, in whole or in part.

**Additional notes**

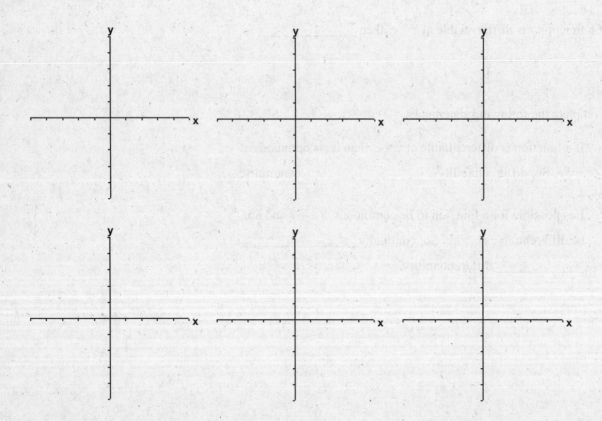

**Homework Assignment**

Page(s)

Exercises

© 2011 Cengage Learning. All Rights Reserved. May not be scanned, copied or duplicated, or posted to a publicly accessible website, in whole or in part.

## Section 3.2  Basic Differentiation Rules and Rates of Change

Course Number

Instructor

Date

**Objective:**    In this lesson you learned how to find the derivative of a
function using basic differentiation rules.

### I. The Constant Rule  (Page 127)

The derivative of a constant function is _____.

If $c$ is a real number, then $\dfrac{d}{dx}[c] = $ _____.

*What you should learn*
How to find the
derivative of a function
using the Constant Rule

### II. The Power Rule  (Pages 128–129)

The **Power Rule** states that if $n$ is a real number, then the
function $f(x) = x^n$ is differentiable and

$\dfrac{d}{dx}\big[x^n\big] = $ _____. For $f$ to be differentiable at

$x = 0$, $n$ must be a number such that $x^{n-1}$ is _____

_____.

*What you should learn*
How to find the
derivative of a function
using the Power Rule

Also, $\dfrac{d}{dx}[x] = $ _____.

**Example 1:**   Find the derivative of the function $f(x) = \dfrac{1}{x^3}$.

**Example 2:**   Find the slope of the graph of $f(x) = x^5$ at $x = 2$.

### III. The Constant Multiple Rule  (Pages 130–131)

The **Constant Multiple Rule** states that if $f$ is a differentiable

function and $c$ is a real number then $cf$ is also differentiable and

$\dfrac{d}{dx}[cf(x)] = $ _____.

*What you should learn*
How to find the
derivative of a function
using the Constant
Multiple Rule

Informally, the Constant Multiple Rule states that _____

_____

_____.

© 2011 Cengage Learning. All Rights Reserved. May not be scanned, copied or duplicated, or posted to a publicly accessible website, in whole or in part.

**Example 3:**  Find the derivative of $f(x) = \dfrac{2x}{5}$

The Constant Multiple Rule and the Power Rule can be combined into one rule. The combination rule is

$$\frac{d}{dx}\left[cx^n\right] = \underline{\hspace{4cm}}.$$

**Example 4:**  Find the derivative of $y = \dfrac{2}{5x^5}$

## IV.  The Sum and Difference Rules  (Page 131)

The **Sum and Difference Rules** of Differentiation state that the sum (or difference) of two differentiable functions $f$ and $g$ is itself differentiable. Moreover, the derivative of $f + g$ (or $f - g$) is the sum (or difference) of the derivatives of $f$ and $g$.

That is, $\dfrac{d}{dx}\left[f(x) + g(x)\right] = \underline{\hspace{5cm}}$

and $\dfrac{d}{dx}\left[f(x) - g(x)\right] = \underline{\hspace{5cm}}$

**Example 5:**  Find the derivative of $f(x) = 2x^3 - 4x^2 + 3x - 1$

| **What you should learn** |
|---|
| How to find the derivative of a function using the Sum and Difference Rules |

## V.  Derivatives of Sine and Cosine Functions  (Page 132)

$$\frac{d}{dx}\left[\sin x\right] = \underline{\hspace{3cm}}$$

$$\frac{d}{dx}\left[\cos x\right] = \underline{\hspace{3cm}}$$

| **What you should learn** |
|---|
| How to find the derivative of the sine function and of the cosine function |

**Example 6:**  Differentiate the function $y = x^2 - 2\cos x$ .

© 2011 Cengage Learning. All Rights Reserved. May not be scanned, copied or duplicated, or posted to a publicly accessible website, in whole or in part.

## VI. Derivatives of Exponential Functions (Page 133)

$$\frac{d}{dx}\left[e^x\right] = \underline{\hspace{4cm}}$$

**What you should learn**
How to find the
derivative of exponential
functions

Give a graphical interpretation of this result.

**Example 7:**  Differentiate the function $y = e^x + 2\sin x$.

## VII. Rates of Change (Pages 134–135)

The derivative can also be used to determine _____

_____.

**What you should learn**
How to use derivatives to
find rates of change

Give some examples of real-life applications of rates of change.

The function $s$ that gives the position (relative to the origin) of an object as a function of time $t$ is called a _____.

The **average velocity** of an object that is moving in a straight line is found as follows.

Average velocity = —————————— = ———

**Example 8:**  If a ball is dropped from the top of a building that is 200 feet tall, and air resistance is neglected, the height $s$ (in feet) of the ball at time $t$ (in seconds) is given by $s = -16t^2 + 200$. Find the average velocity of the object over the interval [1, 3].

Larson/Edwards  **Calculus: Early Transcendental Functions 5e**  Notetaking Guide

© 2011 Cengage Learning. All Rights Reserved. May not be scanned, copied or duplicated, or posted to a publicly accessible website, in whole or in part.

If $s = s(t)$ is the position function for an object moving along a straight line, the (instantaneous) **velocity** of the object at time $t$ is

$$v(t) = \underline{\hspace{3cm}} = \underline{\hspace{3cm}}.$$

In other words, the velocity function is the \underline{\hspace{2cm}} the position function. Velocity can be \underline{\hspace{3cm}}

\underline{\hspace{3cm}}. The \underline{\hspace{2cm}} of an object is the absolute value of its velocity. Speed cannot be \underline{\hspace{2cm}}.

**Example 9:**  If a ball is dropped from the top of a building that is 200 feet tall, and air resistance is neglected, the height $s$ (in feet) of the ball at time $t$ (in seconds) is given by $s(t) = -16t^2 + 200$. Find the velocity of the ball when $t = 3$.

The position function for a free-falling object (neglecting air resistance) under the influence of gravity can be represented by the equation \underline{\hspace{4cm}}, where $s_0$ is the initial height of the object, $v_0$ is the initial velocity of the object, and $g$ is the acceleration due to gravity. On Earth, the value of $g$ is

\underline{\hspace{6cm}}

\underline{\hspace{6cm}}.

**Homework Assignment**

Page(s)

Exercises

© 2011 Cengage Learning. All Rights Reserved. May not be scanned, copied or duplicated, or posted to a publicly accessible website, in whole or in part.

## Section 3.3  Product and Quotient Rules and Higher-Order Derivatives

Course Number

Instructor

Date

**Objective:**    In this lesson you learned how to find the derivative of a function using the Product Rule and Quotient Rule.

### I.  The Product Rule  (Pages 140–141)

The product of two differentiable functions $f$ and $g$ is itself differentiable. The **Product Rule** states that the derivative of the

$fg$ is equal to _____

_____

_____. That is,

$$\frac{d}{dx}[f(x)g(x)] = f(x)g'(x) + g(x)f'(x).$$

***What you should learn***
How to find the derivative of a function using the Product Rule

**Example 1:**   Find the derivative of $y = (4x^2 + 1)(2x - 3)$.

The Product Rule can be extended to cover products that have more than two factors. For example, if $f$, $g$, and $h$ are differentiable functions of $x$, then

$$\frac{d}{dx}[f(x)g(x)h(x)] = \underline{\hspace{5cm}}$$

Explain the difference between the Constant Multiple Rule and the Product Rule.

© 2011 Cengage Learning. All Rights Reserved. May not be scanned, copied or duplicated, or posted to a publicly accessible website, in whole or in part.

# 48 — Chapter 3 Differentiation

## II. The Quotient Rule (Pages 142–144)

The quotient $f/g$ of two differentiable functions $f$ and $g$ is itself differentiable at all values of $x$ for which $g(x) \neq 0$. The derivative of $f/g$ is given by _____

_____

_____ ,

all divided by _____ .

This is called the _____ , and is given by

$$\frac{d}{dx}\left[\frac{f(x)}{g(x)}\right] = \frac{g(x)f'(x) - f(x)g'(x)}{[g(x)]^2}, \qquad g(x) \neq 0 .$$

**Example 2:** Find the derivative of $y = \dfrac{2x+5}{3x}$.

With the Quotient Rule, it is a good idea to enclose all factors and derivatives _____ and to pay special attention to _____

_____ .

## III. Derivatives of Trigonometric Functions (Pages 144–145)

$\dfrac{d}{dx}[\tan x] = $ _____

$\dfrac{d}{dx}[\cot x] = $ _____

$\dfrac{d}{dx}[\sec x] = $ _____

$\dfrac{d}{dx}[\csc x] = $ _____

**Example 3:** Differentiate the function $f(x) = \sin x \sec x$.

*What you should learn* — How to find the derivative of a function using the Quotient Rule

*What you should learn* — How to find the derivative of a trigonometric function

© 2011 Cengage Learning. All Rights Reserved. May not be scanned, copied or duplicated, or posted to a publicly accessible website, in whole or in part.

**IV.  Higher-Order Derivatives**  (Page 146)

**What you should learn**
How to find a higher-order derivative of a function

The derivative of $f'(x)$ is the second derivative of $f(x)$ and is

denoted by _____. The derivative of $f''(x)$ is the

_____ of $f(x)$ and is denoted by $f'''$.

These are examples of _____ of

$f(x)$.

The following notation is used to denoted the _____

of the function $y = f(x)$:

$$\frac{d^6 y}{dx^6} \quad D_x^6[y] \quad y^{(6)} \quad \frac{d^6}{dx^6}[f(x)] \quad f^{(6)}(x)$$

**Example 4:**  Find $y^{(5)}$ for $y = 2x^7 - x^5$.

**Example 5:**  On the moon, a ball is dropped from a height of 100 feet. Its height $s$ (in feet) above the moon's surface is given by $s = -\frac{27}{10}t^2 + 100$. Find the height, the velocity, and the acceleration of the ball when $t = 5$ seconds.

© 2011 Cengage Learning. All Rights Reserved. May not be scanned, copied or duplicated, or posted to a publicly accessible website, in whole or in part.

**Example 6:**   Find $y'''$ for $y = \sin x$.

**Additional notes**

---

**Homework Assignment**

Page(s)

Exercises

---

© 2011 Cengage Learning. All Rights Reserved. May not be scanned, copied or duplicated, or posted to a publicly accessible website, in whole or in part.

## Section 3.4  The Chain Rule

**Objective:**     In this lesson you learned how to find the derivative of a
function using the Chain Rule and General Power Rule.

Course Number

Instructor

Date

**I. The Chain Rule** (Pages 151–153)

The Chain Rule, one of the most powerful differentiation rules,

deals with _____ functions.

*What you should learn*
How to find the
derivative of a composite
function using the Chain
Rule

Basically, the Chain Rule states that if $y$ changes $dy/du$ times as

fast as $u$, and $u$ changes $du/dx$ times as fast as $x$, then $y$ changes

_____ times as fast as $x$.

The **Chain Rule** states that if $y = f(u)$ is a differentiable

function of $u$, and $u = g(x)$ is a differentiable function of $x$, then

$y = f(g(x))$ is a differentiable function of $x$, and

$\dfrac{dy}{dx} = $ —————— • —————— or, equivalently,

$\dfrac{d}{dx}[f(g(x))] = $ _____ .

When applying the Chain Rule, it is helpful to think of the

composite function $f \circ g$ as having two parts, an *inner part* and

an *outer part*. The Chain Rule tells you that the derivative of

$y = f(u)$ is the derivative of the _____ (at

the inner function $u$) *times* the derivative of the _____

_____ . That is, $y' = $ _____ .

**Example 1:**   Find the derivative of $y = (3x^2 - 2)^5$.

**II. The General Power Rule** (Pages 153–154)

The General Power Rule is a special case of the _____

_____ .

*What you should learn*
How to find the
derivative of a function
using the General Power
Rule

© 2011 Cengage Learning. All Rights Reserved. May not be scanned, copied or duplicated, or posted to a publicly accessible website, in whole or in part.

The General Power Rule states that if $y = [u(x)]^n$, where $u$ is a

differentiable function of $x$ and $n$ is a real number, then

$$\frac{dy}{dx} = \underline{\hspace{5cm}}$$                  or, equivalently,

$$\frac{d}{dx}[u^n] = \underline{\hspace{5cm}}.$$

**Example 2:**  Find the derivative of $y = \dfrac{4}{(2x-1)^3}$.

### III.  Simplifying Derivatives  (Page 155)

**Example 3:**  Find the derivative of $y = \dfrac{3x^2}{(1-x^3)^2}$ and simplify.

**What you should learn**
How to simplify the
derivative of a function
using algebra

### IV.  Transcendental Functions and the Chain Rule
   (Page 156)

Complete each of the following "Chain Rule versions" of the
derivatives of the six trigonometric functions.

**What you should learn**
How to find the
derivative of a
transcendental function
using the Chain Rule

$$\frac{d}{dx}[\sin u] = \underline{\hspace{5cm}}$$

$$\frac{d}{dx}[\cos u] = \underline{\hspace{5cm}}$$

$$\frac{d}{dx}[\tan u] = \underline{\hspace{5cm}}$$

$$\frac{d}{dx}[\cot u] = \underline{\hspace{5cm}}$$

$$\frac{d}{dx}[\sec u] = \underline{\hspace{5cm}}$$

$$\frac{d}{dx}[\csc u] = \underline{\hspace{5cm}}$$

$$\frac{d}{dx}[e^u] = \underline{\hspace{5cm}}$$

© 2011 Cengage Learning. All Rights Reserved. May not be scanned, copied or duplicated, or posted to a publicly accessible website, in whole or in part.

**Example 4:**  Differentiate the function $y = \sec 4x$.

**Example 5:**  Differentiate the function $y = x^2 - \cos(2x + 1)$.

### V.  The Derivative of the Natural Logarithmic Function
   (Pages 157–158)

Let $u$ be a differentiable function of $x$. Complete the following
rules of differentiation for the natural logarithmic function:

$$\frac{d}{dx}[\ln x] = \underline{\hspace{3cm}}, x > 0$$

$$\frac{d}{dx}[\ln u] = \underline{\hspace{4cm}}, u > 0$$

> **What you should learn**
> How to find the derivative of a function involving the natural logarithmic function

**Example 6:**  Find the derivative of $f(x) = x^2 \ln x$.

If $u$ is a differentiable function of $x$ such that $u \neq 0$, then
$$\frac{d}{dx}[\ln|u|] = \underline{\hspace{3cm}}.$$ In other words, functions of
the form $y = \ln|u|$ can be differentiated as if $\underline{\hspace{3cm}}$

$\underline{\hspace{5cm}}$.

### VI.  Bases Other than $e$  (Pages 159–160)

If $a$ is a positive real number $(a \neq 1)$ and $x$ is any real number,

then the **exponential function to the base $a$** is denoted by $a^x$

and is defined by $\underline{\hspace{4cm}}$. If $a = 1$, then

$y = 1^x = 1$ is a $\underline{\hspace{4cm}}$.

> **What you should learn**
> How to define and differentiate exponential functions that have bases other than $e$

Larson/Edwards  **Calculus: Early Transcendental Functions 5e**  Notetaking Guide

© 2011 Cengage Learning. All Rights Reserved. May not be scanned, copied or duplicated, or posted to a publicly accessible website, in whole or in part.

If $a$ is a positive real number $(a \neq 1)$ and $x$ is any positive real number, then the **logarithmic function to the base $a$** is denoted by $\log_a x$ and is defined by $\log_a x =$ _____.

To differentiate exponential and logarithmic functions to other bases, you have two options:

1.

2.

Let $a$ be a positive real number $(a \neq 1)$ and let $u$ be a differentiable function of $x$. Complete the following formulas for the derivatives for bases other than $e$.

$\dfrac{d}{dx}\left[a^x\right] =$ _____.

$\dfrac{d}{dx}\left[a^u\right] =$ _____.

$\dfrac{d}{dx}[\log_a x] =$ _____.

$\dfrac{d}{dx}[\log_a u] =$ _____.

**Homework Assignment**

Page(s)

Exercises

© 2011 Cengage Learning. All Rights Reserved. May not be scanned, copied or duplicated, or posted to a publicly accessible website, in whole or in part.

## Section 3.5  Implicit Differentiation

**Objective:**    In this lesson you learned how to find the derivative of a
function using implicit differentiation.

Course Number

Instructor

Date

### I.  Implicit and Explicit Functions  (Page 166)

Up to this point in the text, most functions have been expressed
in **explicit form** $y = f(x)$, meaning that _____
_____. However,
some functions are only _____ by an equation.

Give an example of a function in which $y$ is **implicitly** defined as a
function of $x$.

*What you should learn*
How to distinguish
between functions written
in implicit form and
explicit form

**Implicit differentiation** is a procedure for taking the derivative
of an implicit function when you are unable to _____
_____.

To understand how to find $\dfrac{dy}{dx}$ implicitly, realize that the

differentiation is taking place _____. This
means that when you differentiate terms involving $x$ alone, _____
_____. However, when you
differentiate terms involving $y$, you must apply _____
_____ because you are assuming that $y$ is defined
_____ as a differentiable function of $x$.

**Example 1:**  Differentiate the expression with respect to $x$:
$4x + y^2$

© 2011 Cengage Learning. All Rights Reserved. May not be scanned, copied or duplicated, or posted to a publicly accessible website, in whole or in part.

## II. Implicit Differentiation (Pages 167–170)

Consider an equation involving $x$ and $y$ in which $y$ is a differentiable function of $x$. List the four guidelines for applying implicit differentiation to find $dy/dx$.

*What you should learn*
How to use implicit differentiation to find the derivative of a function

1.

2.

3.

4.

**Example 2:**  Find $dy/dx$ for the equation $4y^2 - x^2 = 1$.

## III. Logarithmic Differentiation (Page 171)

Describe the procedure for logarithmic differentiation.

*What you should learn*
How to find derivatives of functions using logarithmic differentiation

**Homework Assignment**

Page(s)

Exercises

© 2011 Cengage Learning. All Rights Reserved. May not be scanned, copied or duplicated, or posted to a publicly accessible website, in whole or in part.

## Section 3.6  Derivatives of Inverse Functions

**Objective:**    In this lesson you learned how to find the derivative of
an inverse function.

Course Number

Instructor

Date

**I. Derivative of an Inverse Function**  (Pages 175–176)

Let $f$ be a function whose domain is an interval $I$. If $f$ has an
inverse function, then the following statements are true.

1.

2.

**What you should learn**
How to find the
derivative of an inverse
function

Let $f$ be a function that is differentiable on an interval $I$. If $f$ has
an inverse function $g$, then $g$ is _____
_____ . Moreover,

$$g'(x) = \frac{1}{f'(g(x))}, \quad f'(g(x)) \neq 0.$$

This last theorem can be interpreted to mean that the graphs of

the inverse functions $f$ and $f^{-1}$ have _____

_____ at points $(a, b)$ and $(b, a)$.

**II. Derivatives of Inverse Trigonometric Functions**
    (Pages 176–178)

Let $u$ be a differentiable function of $x$.

**What you should learn**
How to differentiate an
inverse trigonometric
function

$$\frac{d}{dx}[\arcsin u] = \frac{\rule{2cm}{0.4pt}}{\sqrt{\rule{1.5cm}{0pt}}}$$

$$\frac{d}{dx}[\arccos u] = \frac{\rule{2cm}{0.4pt}}{\sqrt{\rule{1.5cm}{0pt}}}$$

$$\frac{d}{dx}[\arctan u] = \rule{2cm}{0.4pt}$$

© 2011 Cengage Learning. All Rights Reserved. May not be scanned, copied or duplicated, or posted to a publicly accessible website, in whole or in part.

$$\frac{d}{dx}\left[\text{arc cot}\, u\right] = \underline{\hspace{3cm}}$$

$$\frac{d}{dx}\left[\text{arc sec}\, u\right] = \frac{\underline{\hspace{2cm}}}{\sqrt{\underline{\hspace{2cm}}}}$$

$$\frac{d}{dx}\left[\text{arc csc}\, u\right] = \frac{\underline{\hspace{2cm}}}{\sqrt{\underline{\hspace{2cm}}}}$$

You should notice that the derivatives of $\text{arccos}\, u$, $\text{arc cot}\, u$, and

$\text{arc csc}\, u$ are the _____ of the derivatives

of $\text{arcsin}\, u$, $\text{arctan}\, u$, and $\text{arcsec}\, u$, respectively.


### III. Review of Basic Differentiation Rules  (Pages 178–179)

An elementary function is _____

_____

_____.

The algebraic functions are _____

_____.

The transcendental functions are _____

_____.

<table>
<tr><td>What you should learn<br>Review the basic<br>differentiation formulas<br>for elementary functions</td></tr>
</table>

---

**Homework Assignment**

Page(s)

Exercises

© 2011 Cengage Learning. All Rights Reserved. May not be scanned, copied or duplicated, or posted to a publicly accessible website, in whole or in part.

## Section 3.7  Related Rates

**Objective:**      In this lesson you learned how to find a related rate.

Course Number

Instructor

Date

### I. Finding Related Rates  (Page 182)

Another important use of the Chain Rule is to find the rates of change of two or more related variables that are changing with respect to _____.

*What you should learn*
How to find a related rate

**Example 1:**    The variables $x$ and $y$ are differentiable functions of $t$ and are related by the equation $y = 2x^3 - x + 4$. When $x = 2$, $dx/dt = -1$. Find $dy/dt$ when $x = 2$.

### II. Problem Solving with Related Rates  (Pages 183–186)

List the guidelines for solving a related-rate problems.

1.

2.

3.

4.

*What you should learn*
How to use related rates to solve real-life problems

**Example 2:**    Write a mathematical model for the following related-rate problem situation:
The population of a city is decreasing at the rate of 100 people per month.

© 2011 Cengage Learning. All Rights Reserved. May not be scanned, copied or duplicated, or posted to a publicly accessible website, in whole or in part.

**Additional notes**

---

**Homework Assignment**

Page(s)

Exercises

---

© 2011 Cengage Learning. All Rights Reserved. May not be scanned, copied or duplicated, or posted to a publicly accessible website, in whole or in part.

## Section 3.8   Newton's Method

**Objective:**      In this lesson you learned how to approximate a zero of
a function using Newton's Method.

Course Number

Instructor

Date

**I. Newton's Method**   (Pages 191–194)

**Newton's Method** is _____

_____, and it uses _____

_____

_____.

*What you should learn*
How to approximate a
zero of a function using
Newton's Method

Let $f(c) = 0$, where $f$ is differentiable on an open interval
containing $c$. To use Newton's Method to approximate $c$, use the
following steps.

1.

2.

3.

Each successive application of this procedure is called an

_____.

When the approximations given by Newton's Method approach a

limit, the sequence $x_1, x_2, x_3, \ldots, x_n, \ldots$ is said to

_____. Moreover, if the limit is $c$, it

can be shown that $c$ must be _____.

Newton's Method does not always yield a convergent sequence.

One way it can fail to do so is if _____

or if _____.

© 2011 Cengage Learning. All Rights Reserved. May not be scanned, copied or duplicated, or posted to a publicly accessible website, in whole or in part.

When the first situation is encountered, it can usually be

overcome by _____.


**Additional notes**

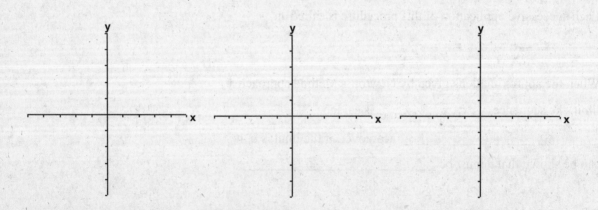

---

**Homework Assignment**

Page(s)

Exercises

---

© 2011 Cengage Learning. All Rights Reserved. May not be scanned, copied or duplicated, or posted to a publicly accessible website, in whole or in part.

# Chapter 4    Applications of Differentiation

Course Number

Instructor

Date

## Section 4.1  Extrema on an Interval

**Objective:** In this lesson you learned how to use a derivative to locate the minimum and maximum values of a function on a closed interval.

---

**Important Vocabulary**         Define each term or concept.

**Relative maximum**


**Relative minimum**


**Critical number**

---

### I.  Extrema of a Function  (Page 204)

Let $f$ be defined on an interval $I$ containing $c$.

1.  $f(c)$ is the **minimum of $f$ on $I$** if _____

_____.

2.  $f(c)$ is the **maximum of $f$ on $I$** if _____

_____.

The minimum and maximum of a function on an interval are the

_____, or extrema (the singular

form of extrema is _____), of the function on the

interval. The minimum and maximum of a function on an

interval are also called the _____

_____, or the _____

_____, on the interval.

The **Extreme Value Theorem** states that if $f$ is continuous on a

closed interval $[a, b]$, then _____

_____.

**What you should learn**
How to understand the definition of extrema of a function on an interval

© 2011 Cengage Learning. All Rights Reserved. May not be scanned, copied or duplicated, or posted to a publicly accessible website, in whole or in part.

## II.  Relative Extrema and Critical Numbers  (Pages 205–206)

If $f$ has a relative minimum or relative maximum when $x = c$, then $c$ is a _____ of $f$.

**What you should learn**
How to understand the definition of relative extrema of a function on an open interval

## III.  Finding Extrema on a Closed Interval  (Pages 207–208)

To find the extrema of a continuous function $f$ on a closed interval $[a, b]$, use the following steps.

**What you should learn**
How to find extrema on a closed interval

1.

2.

3.

4.

**Example 1:**   Find the extrema of the function
$$f(x) = x^3 + 6x^2 - 15x + 2 \text{ on the interval } [-6, 6].$$

The critical numbers of a function need not produce _____ _____.

---

**Homework Assignment**

Page(s)

Exercises

---

© 2011 Cengage Learning. All Rights Reserved. May not be scanned, copied or duplicated, or posted to a publicly accessible website, in whole or in part.

## Section 4.2   Rolle's Theorem and the Mean Value Theorem

**Objective:**        In this lesson you learned how numerous results in this chapter depend on two important theorems called *Rolle's Theorem* and the *Mean Value Theorem*.

Course Number

Instructor

Date

### I. Rolle's Theorem  (Pages 212–213)

The Extreme Value Theorem states that a continuous function on a closed interval [a, b] must have _____ _____. Both of these values, however, can occur at _____.

**Rolle's Theorem** gives conditions that guarantee the existence of an extreme value in _____ _____.

*What you should learn*
How to understand and use Rolle's Theorem

The statement of Rolle's Theorem says:    Let $f$ be continuous on the closed interval [a, b] and differentiable on the open interval (a, b). If $f(a) = f(b)$, then there is _____ _____.

If $f$ satisfies the conditions of Rolle's Theorem, then there must be at least one $x$-value between $a$ and $b$ at which the graph of $f$ has _____.

Alternatively, Rolle's Theorem states that if $f$ satisfies the conditions of the theorem, there must be at least one point between $a$ and $b$ at which the derivative is _____.

### II. The Mean Value Theorem  (Pages 214–215)

The **Mean Value Theorem** states that if $f$ is continuous on _____ and differentiable on _____, then there exists _____ such that

$$f'(c) = \frac{f(b) - f(a)}{b - a}.$$

*What you should learn*
How to understand and use the Mean Value Theorem

© 2011 Cengage Learning. All Rights Reserved. May not be scanned, copied or duplicated, or posted to a publicly accessible website, in whole or in part.

The Mean Value Theorem has implications for both basic

interpretations of the derivative. Geometrically, the theorem

guarantees  the existence of _____

_____

_____. In terms of rates of change,

the Mean Value Theorem implies that there must be _____

_____

_____

_____.


A useful alternative form of the Mean Value Theorem is as

follows:  If $f$ is continuous on $[a, b]$ and differentiable on $(a, b)$,

then there exists a number $c$ in $(a, b)$ such that

_____.


**Homework Assignment**

Page(s)

Exercises

© 2011 Cengage Learning. All Rights Reserved. May not be scanned, copied or duplicated, or posted to a publicly accessible website, in whole or in part.

## Section 4.3    Increasing and Decreasing Functions and the First Derivative Test

Course Number

Instructor

Date

**Objective:**    In this lesson you learned how to use the first derivative to determine whether a function is increasing or decreasing.

---

**Important Vocabulary**          Define each term or concept.

**Increasing function**

**Decreasing function**

---

**I. Increasing and Decreasing Functions**  (Pages 219–220)

A function is **increasing** if, as $x$ moves _____,

its graph moves _____. A function is **decreasing** if its

graph moves _____ as $x$ moves _____.

> **What you should learn**
> How to determine intervals on which a function is increasing or decreasing

Let $f$ be a function that is continuous on the closed interval $[a, b]$ and differentiable on the open interval $(a, b)$.

If $f'(x) > 0$ for all $x$ in $(a, b)$, then $f$ is _____ on $[a, b]$.

If $f'(x) < 0$ for all $x$ in $(a, b)$, then $f$ is _____ on $[a, b]$.

If $f'(x) = 0$ for all $x$ in $(a, b)$, then $f$ is _____ on $[a, b]$.

The first of these tests for increasing and decreasing functions

can be interpreted as follows:  if the first derivative of a function

is positive for all values of $x$ in an interval, then the function is

_____ on that interval.

Interpret the other two tests in a similar way.

© 2011 Cengage Learning. All Rights Reserved. May not be scanned, copied or duplicated, or posted to a publicly accessible website, in whole or in part.

**Example 1:**   Find the open intervals on which the function is increasing
or decreasing:  $f(x) = -x^2 + 10x - 21$

Let $f$ be a continuous function on the interval $(a, b)$. List the
steps for finding the intervals on which $f$ is increasing or
decreasing.

1.

2.

3.

A function is **strictly monotonic** on an interval if _____

_____

_____.

**II.  The First Derivative Test**  (Pages 221–225)

Let $c$ be a critical number of a function $f$ that is continuous on an
open interval $I$ containing $c$. The **First Derivative Test** states
that if $f$ is differentiable on the interval (except possibly at $c$),
then $f(c)$ can be classified as follows:

**What you should learn**
How to apply the First
Derivative Test to find
relative extrema of a
function

1.  If $f'(x)$ changes from negative to positive at $c$, then $f$

    has a _____ at $(c, f(c))$.

2.  If $f'(x)$ changes from positive to negative at $c$, then $f$

    has a _____ at $(c, f(c))$.

3.  If $f'(x)$ is positive on both sides of $c$ or negative on

    both sides of $c$, then $f(c)$ is _____

    _____.

© 2011 Cengage Learning. All Rights Reserved. May not be scanned, copied or duplicated, or posted to a publicly accessible website, in whole or in part.

In your own words, describe how to find the relative extrema of
a function $f$.

**Example 2:**    Find all relative extrema of the function
$$f(x) = x^3 - 7x^2 - 38x + 240.$$

© 2011 Cengage Learning. All Rights Reserved. May not be scanned, copied or duplicated, or posted to a publicly accessible website, in whole or in part.

**Additional notes**

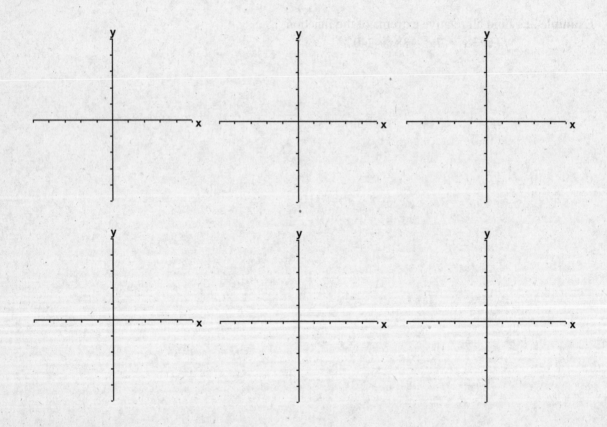

**Homework Assignment**

Page(s)

Exercises

© 2011 Cengage Learning. All Rights Reserved. May not be scanned, copied or duplicated, or posted to a publicly accessible website, in whole or in part.

## Section 4.4  Concavity and the Second Derivative Test

**Objective:**    In this lesson you learned how to use the second
derivative to determine whether the graph of a function
is concave upward or concave downward.

Course Number

Instructor

Date

---

**Important Vocabulary**          Define each term or concept.

**Concave upward**

**Concave downward**

**Point of inflection**

---

**I. Concavity**  (Pages 230–232)

Let $f$ be differentiable on an open interval $I$. If the graph of $f$ is
**concave upward**, then the graph of $f$ lies _____ all of its
tangent lines on $I$.

Let $f$ be differentiable on an open interval $I$. If the graph of $f$ is
**concave downward,** then the graph of $f$ lies _____ all
of its tangent lines on $I$.

As a test for concavity, let $f$ be a function whose second
derivative exists on an open interval $I$.

  1. If $f''(x) > 0$ for all $x$ in $I$, then the graph of $f$ is _____
      _____ in $I$.

  2. If $f''(x) < 0$ for all $x$ in $I$, then the graph of $f$ is _____
      _____ in $I$.

In your own words, describe how to apply the Concavity Test.

---

*What you should learn*
How to determine
intervals on which a
function is concave
upward or concave
downward

---

© 2011 Cengage Learning. All Rights Reserved. May not be scanned, copied or duplicated, or posted to a publicly accessible website, in whole or in part.

**Example 1:**   Describe the concavity of the function
$$f(x) = 1 - 3x^2 .$$

## II. Points of Inflection  (Pages 232–233)

To locate possible points of inflection, you can determine

_____

_____.

> **What you should learn**
> How to find any points of inflection of the graph of a function

State Theorem 4.8 for Points of Inflection.

**Example 2:**   Find the points of inflection of
$$f(x) = -\frac{1}{2}x^4 + 10x^3 - 48x^2 + 4 .$$

The converse of Theorem 4.8 is _____.

That is, it is possible for the second derivative to be 0 at a point

that is _____.

## III. The Second-Derivative Test  (Page 234)

Let be a function such that $f'(c) = 0$ and the second derivative
of $f$ exists on an open interval containing $c$. Then the **Second-Derivative Test** states:

> **What you should learn**
> How to apply the Second Derivative Test to find relative extrema of a function

1.

2.

> **Homework Assignment**
>
> Page(s)
>
> Exercises

© 2011 Cengage Learning. All Rights Reserved. May not be scanned, copied or duplicated, or posted to a publicly accessible website, in whole or in part.

## Section 4.5  Limits at Infinity

**Objective:**    In this lesson you learned how to find horizontal
                  asymptotes of the graph of a function.

| Course Number |
| Instructor |
| Date |

**I. Limits at Infinity**  (Page 238)

*What you should learn*
How to determine (finite)
limits at infinity

To say that a statement is true as $x$ increases without bound
means that for some (large) real number $M$, the statement is true
for all $x$ in the interval _____.

Let $L$ be a real number. The definition of **limit at infinity** states
that

1.  $\lim\limits_{x \to \infty} f(x) = L$ means _____

    _____

    _____.

2.  $\lim\limits_{x \to -\infty} f(x) = L$ means _____

    _____

    _____.

**II. Horizontal Asymptotes**  (Pages 239–243)

*What you should learn*
How to determine the
horizontal asymptotes, if
any, of the graph of a
function

The line $y = L$ is a _____ of the graph
of $f$ if $\lim\limits_{x \to -\infty} f(x) = L$ or $\lim\limits_{x \to \infty} f(x) = L$.

Notice that from this definition, if follows that the graph of a
function of $x$ can have at most _____

_____.

If $r$ is a positive rational number and $c$ is any real number, then

$\lim\limits_{x \to \infty} \dfrac{c}{x^r} =$ _____. Furthermore, if $x^r$ is defined when

$x < 0$, then $\lim\limits_{x \to -\infty} \dfrac{c}{x^r} =$ _____.

$\lim\limits_{x \to -\infty} e^x =$ _____ and $\lim\limits_{x \to \infty} e^{-x} =$ _____.

© 2011 Cengage Learning. All Rights Reserved. May not be scanned, copied or duplicated, or posted to a publicly accessible website, in whole or in part.

**Example 1:**   Find the limit:  $\displaystyle\lim_{x \to \infty}\left(2 + \frac{3}{x^2}\right)$

If an **indeterminate form** _____ is encountered

while finding a limit at infinity, you can resolve this problem by

_____

_____.

Complete the following guidelines for finding limits at $\pm\,\infty$ of
rational functions.

1. _____

   _____

   _____.

2. _____

   _____

   _____.

3. _____

   _____

   _____.

**Example 2:**   Find the limit:  $\displaystyle\lim_{x \to \infty}\frac{x^3 - 1}{1 - 13x + 2x^2 - 5x^3}$

### III.  Infinite Limits at Infinity  (Page 244)

Many functions do not approach a finite limit as $x$ increases (or

decreases) without bound. _____ are

one type of function that does not have a finite limit at infinity.

> **What you should learn**
> How to determine infinite
> limits at infinity

© 2011 Cengage Learning. All Rights Reserved. May not be scanned, copied or duplicated, or posted to a publicly accessible website, in whole or in part.

Let $f$ be a function defined on the interval $(a, \infty)$. The definition
of **infinite limits at infinity** states that

    1.  $\lim\limits_{x \to \infty} f(x) = \infty$ means _____

    _____

    _____ .

    2.  $\lim\limits_{x \to \infty} f(x) = -\infty$ means _____

    _____

    _____ .

**Example 3:**   Find the limit:   $\lim\limits_{x \to \infty}(2x^2 - 9x + 1)$.

© 2011 Cengage Learning. All Rights Reserved. May not be scanned, copied or duplicated, or posted to a publicly accessible website, in whole or in part.

**Additional notes**

**Homework Assignment**

Page(s)

Exercises

© 2011 Cengage Learning. All Rights Reserved. May not be scanned, copied or duplicated, or posted to a publicly accessible website, in whole or in part.

## Section 4.6  A Summary of Curve Sketching

**Objective:**      In this lesson you learned how to graph a function using
                    the techniques from Chapters 1–4.

Course Number

Instructor

Date

**I.  Analyzing the Graph of a Function**  (Pages 249–255)

List some of the concepts that you have studied thus far that are
useful in analyzing the graph of a function.

*What you should learn*
How to analyze and
sketch the graph of a
function

List three guidelines for analyzing the graph of a function.

1.

2.

3.

The graph of a rational function (having no common factors and
whose denominator is of degree 1 or greater) has a _____
_____ if the degree of the numerator exceeds the
degree of the denominator by exactly 1.

To find the slant asymptote, _____

_____

_____.

© 2011 Cengage Learning. All Rights Reserved. May not be scanned, copied or duplicated, or posted to a publicly accessible website, in whole or in part.

In general, a polynomial function of degree $n$ can have at most

_____ relative extrema, and at most

_____ points of inflection. Moreover,

polynomial functions of even degree must have _____

_____ relative extremum.

**Homework Assignment**

Page(s)

Exercises

© 2011 Cengage Learning. All Rights Reserved. May not be scanned, copied or duplicated, or posted to a publicly accessible website, in whole or in part.

## Section 4.7  Optimization Problems

**Objective:**       In this lesson you learned how to solve optimization
                  problems.

Course Number

Instructor

Date

**I. Applied Minimum and Maximum Problems**
   (Pages 259–264)

What does "optimization problem" mean?

*What you should learn*
How to solve applied
minimum and maximum
problems

In an optimization problem, the **primary equation** is one that
_____.

The feasible domain of a function consists of _____
_____.

A secondary equation is used to _____
_____.

List the steps for solving optimization problems.

1.

2.

3.

4.

5.

© 2011 Cengage Learning. All Rights Reserved. May not be scanned, copied or duplicated, or posted to a publicly accessible website, in whole or in part.

**Additional notes**

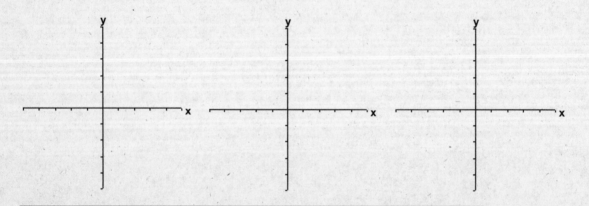

**Homework Assignment**

Page(s)

Exercises

© 2011 Cengage Learning. All Rights Reserved. May not be scanned, copied or duplicated, or posted to a publicly accessible website, in whole or in part.

## Section 4.8   Differentials

**Objective:**        In this lesson you learned how to use approximation
                      techniques to solve problems.

Course Number

Instructor

Date

---

**Important Vocabulary**            Define each term or concept.

**Differential of $x$**

**Differential of $y$**

---

### I. Tangent Line Approximations   (Page 271)

Consider a function $f$ that is differentiable at $c$. The equation for
the tangent line at the point $(c, f(c))$ is given by

_____, and is

called _____

_____. Because $c$ is a

constant, $y$ is a _____ of $x$. Moreover, by

restricting the values of $x$ to be sufficiently close to $c$, the values

of $y$ can be used as approximations (to any desired accuracy) of

_____. In other words, as

$x \to c$, the limit of $y$ is _____.

*What you should learn*
How to understand the
concept of a tangent line
approximation

### II. Differentials   (Page 272)

When the tangent line to the graph of $f$ at the point $(c, f(c))$ is

used as an approximation of the graph of $f$, the quantity $x - c$ is

called the _____ and is denoted by

_____. When $\Delta x$ is small, the change in $y$ (denoted

by $\Delta y$) can be approximated as _____.

For such an approximation, the quantity $\Delta x$ is traditionally

denoted by _____ and is called the **differential of $x$.**

The expression $f'(x)dx$ is denoted by _____, and is

called the **differential of $y$.**

*What you should learn*
How to compare the
value of the differential,
$dy$, with the actual
change in $y$, $\Delta y$

Larson/Edwards   **Calculus: Early Transcendental Functions 5e**   Notetaking Guide

© 2011 Cengage Learning. All Rights Reserved. May not be scanned, copied or duplicated, or posted to a publicly accessible website, in whole or in part.

In many types of applications, the differential of $y$ can be used as

_____. That

is, $\Delta y \approx$ _____ or $\Delta y \approx$ _____.


### III. Error Propagation   (Page 273)

Physicists and engineers tend to make liberal use of the

approximation of $\Delta y$ by $dy$. One way this occurs in practice is in

the _____

_____. For example, if you let $x$ represent the

measured value of a variable and let $x + \Delta x$ represent the exact

value, then $\Delta x$ is _____.

Finally, if the measured value $x$ is used to compute another value

$f(x)$, the difference between $f(x + \Delta x)$ and $f(x)$ is the _____

_____.

> **What you should learn**
> How to estimate a propagated error using a differential

### IV. Calculating Differentials   (Pages 274–275)

Each of the differentiation rules that you studied in Chapter 3

can be written in _____.

Suppose $u$ and $v$ are differentiable functions of $x$. By the

definition of differentials, you have

$du =$ _____ and $dv =$ _____

Complete the following differential forms of common
differentiation rules:

Constant Multiple Rule: _____

Sum or Difference Rule: _____

Product Rule: _____

Quotient Rule: _____

> **What you should learn**
> How to find the differential of a function using differentiation formulas

> **Homework Assignment**
>
> Page(s)
>
> Exercises

© 2011 Cengage Learning. All Rights Reserved. May not be scanned, copied or duplicated, or posted to a publicly accessible website, in whole or in part.

# Chapter 5    Integration

## Section 5.1  Antiderivatives and Indefinite Integration

| Course Number |
| Instructor |
| Date |

**Objective:** In this lesson you learned how to evaluate indefinite integrals using basic integration rules.

---

**Important Vocabulary**    Define each term or concept.

**Antiderivative**

---

### I. Antiderivatives  (Pages 284–285)

If $F$ is an antiderivative of $f$ on an interval $I$, then $G$ is an antiderivative of $f$ on the interval $I$ if and only if $G$ is of the form _____, for all $x$ in $I$ where $C$ is a constant. The entire family of antiderivatives of a function can be represented by _____
_____. The constant $C$ is called the _____. The family of functions represented by $G$ is the _____.

A **differential equation** in $x$ and $y$ is an equation that _____
_____.

Give an example of a *differential equation* and its **general solution.**

> **What you should learn**
> How to write the general solution of a differential equation

### II. Notation for Antiderivatives  (Page 285)

The operation of finding all solutions of the equation $dy = f(x)\,dx$ is called _____

_____ and is denoted by the symbol $\int$ , which

is called an _____.

> **What you should learn**
> How to use indefinite integral notation for antiderivatives

© 2011 Cengage Learning. All Rights Reserved. May not be scanned, copied or duplicated, or posted to a publicly accessible website, in whole or in part.

Use the terms *antiderivative, constant of integration, differential, integral sign,* and *integrand* to label the following notation:

$$\int f(x)\,dx = F(x) + C$$

The differential in the indefinite integral identifies _____

_____.

The notation $\int f(x)\,dx = F(x) + C$, where $C$ is an arbitrary

constant, means that $F$ is _____

_____.

### III. Basic Integration Rules (Pages 286–288)

Complete the following basic integration rules, which follow from differentiation formulas.

**What you should learn**
How to use basic integration rules to find antiderivatives

1.  $\int k\,dx = $ _____.

2.  $\int kf(x)\,dx = $ _____

3.  $\int [f(x) + g(x)]\,dx = $ _____

4.  $\int [f(x) - g(x)]\,dx = $ _____

5.  _____ $= \dfrac{x^{n+1}}{n+1} + C, \quad n \neq -1$

6.  $\int 0\,dx = $ _____

7.  $\int \cos x\,dx = $ _____

8.  $\int \sin x\,dx = $ _____

© 2011 Cengage Learning. All Rights Reserved. May not be scanned, copied or duplicated, or posted to a publicly accessible website, in whole or in part.

9.  $\int \sec^2 x \, dx = $ _____

10. $\int \sec x \tan x \, dx = $ _____

11. $\int \csc^2 x \, dx = $ _____

12. $\int \csc x \cot x \, dx = $ _____

13. $\int e^x \, dx = $ _____

14. _____ $= \left( \dfrac{1}{\ln a} \right) a^x + C$

15. $\int \dfrac{1}{x} \, dx = $ _____

**Example 1:**  Find $\int -3 \, dx$ .

**Example 2:**  Find $\int 2x^2 \, dx$ .

**Example 3:**  Find $\int (1 - 2x) \, dx$ .

**IV. Initial Conditions and Particular Solutions**
    (Pages 289–291)

The equation $y = \int f(x) \, dx$ has many solution, each differing

from the others _____. This means that the

graphs of any two antiderivatives of $f$ are _____

_____.

> **What you should learn**
> How to find a particular
> solution of a differential
> equation

Larson/Edwards  **Calculus: Early Transcendental Functions 5e**  Notetaking Guide

© 2011 Cengage Learning. All Rights Reserved. May not be scanned, copied or duplicated, or posted to a publicly accessible website, in whole or in part.

In many applications of integration, you are given enough

information to determine a _____.

To do this, you need only know the value of $y = F(x)$ for one

value of $x$, called an _____.


**Example 4:**   Solve the differential equation $\dfrac{dC}{dx} = -0.2x + 40$,

                where $C(180) = 89.90$.


**Homework Assignment**

Page(s)

Exercises

© 2011 Cengage Learning. All Rights Reserved. May not be scanned, copied or duplicated, or posted to a publicly accessible website, in whole or in part.

## Section 5.2  Area

**Objective:**     In this lesson you learned how to evaluate a sum and
approximate the area of a plane region.

Course Number

Instructor

Date

### I. Sigma Notation  (Pages 295–296)

The sum of $n$ terms $a_1, a_2, a_3, \ldots, a_n$ is written as

$$\sum_{i=1}^{n} a_i = a_1 + a_2 + a_3 + \cdots + a_n \text{, where } i \text{ is the } \underline{\hspace{3cm}}$$

$\underline{\hspace{3cm}}$, $a_i$ is the $\underline{\hspace{4cm}}$,

and $n$ and $1$ are the $\underline{\hspace{5cm}}$

$\underline{\hspace{3cm}}$.

**What you should learn**
How to use sigma
notation to write and
evaluate a sum

Complete the following properties of summation which are
derived using the associative and commutative properties of
addition and the distributive property of addition over
multiplication.

$$\sum_{i=1}^{n} ka_i = \underline{\hspace{4cm}}$$

$$\sum_{i=1}^{n} (a_i \pm b_i) = \underline{\hspace{4cm}}$$

Now complete the following summation formulas.

1. $\displaystyle\sum_{i=1}^{n} c = \underline{\hspace{3cm}}$

2. $\displaystyle\sum_{i=1}^{n} i = \underline{\hspace{3cm}}$

3. $\displaystyle\sum_{i=1}^{n} i^2 = \underline{\hspace{3cm}}$

4. $\displaystyle\sum_{i=1}^{n} i^3 = \underline{\hspace{3cm}}$

© 2011 Cengage Learning. All Rights Reserved. May not be scanned, copied or duplicated, or posted to a publicly accessible website, in whole or in part.

## II. Area  (Page 297)

In your own words, explain the exhaustion method that the
ancient Greeks used to determine formulas for the areas of
general regions.

**What you should learn**
How to understand the
concept of area

## III. Area of a Plane Region  (Page 298)

Describe how to approximate the area of a plane region.

**What you should learn**
How to approximate the
area of a plane region

## IV. Upper and Lower Sums  (Pages 299–303)

Consider a plane region bounded above by the graph of a
nonnegative, continuous function $y = f(x)$. The region is

bounded below by the _____, and the left and right

boundaries of the region are the vertical lines $x = a$ and $x = b$.

To approximate the area of the region, begin by _____

_____, each of width

_____. Because $f$ is continuous, the

Extreme Value Theorem guarantees the existence of a

_____ in

each subinterval. The value $f(m_i)$ is _____

**What you should learn**
How to find the area of a
plane region using limits

© 2011 Cengage Learning. All Rights Reserved. May not be scanned, copied or duplicated, or posted to a publicly accessible website, in whole or in part.

_____ and the value of $f(M_i)$

is _____.

An **inscribed rectangle** _____ the $i$th subregion

and a **circumscribed rectangle** _____ the

$i$th subregion. The height of the $i$th inscribed rectangle is

_____ and the height of the $i$th circumscribed

rectangle is _____ . For each $i$, the area of the

inscribed rectangle is _____ the area

of the circumscribed rectangle. The sum of the areas of the

inscribed rectangles is called _____, and the

sum of the areas of the circumscribed rectangles is called

_____.

$$\text{_____} = s(n) = \sum_{i=1}^{n} f(m_i)\Delta x$$

$$\text{_____} = S(n) = \sum_{i=1}^{n} f(M_i)\Delta x$$

The actual area of the region lies between _____

_____.

Let $f$ be continuous and nonnegative on the interval $[a, b]$. The

limits as $n \to \infty$ of both the lower and upper sums exist and are

_____ . That is,

$$\lim_{n\to\infty} s(n) = \lim_{n\to\infty} \sum_{i=1}^{n} f(m_i)\Delta x$$

$$= \lim_{n\to\infty} \sum_{i=1}^{n} f(M_i)\Delta x$$

$$= \lim_{n\to\infty} S(n)$$

where $\Delta x = (b-a)/n$ and $f(m_i)$ and $f(M_i)$ are the minimum
and maximum values of $f$ on the subinterval.

© 2011 Cengage Learning. All Rights Reserved. May not be scanned, copied or duplicated, or posted to a publicly accessible website, in whole or in part.

**Definition of the Area of a Region in the Plane**

Let $f$ be continuous and nonnegative on the interval $[a, b]$. The
area of the region bounded by the graph of $f$, the $x$-axis, and the
vertical lines $x = a$ and $x = b$ is

$$\text{Area} = \lim_{n \to \infty} \sum_{i=1}^{n} \underline{\hspace{6cm}} \quad , \quad x_{i-1} \leq c_i \leq x_i$$

where $\Delta x = (b - a) / n$.

---

**Homework Assignment**

Page(s)

Exercises

---

© 2011 Cengage Learning. All Rights Reserved. May not be scanned, copied or duplicated, or posted to a publicly accessible website, in whole or in part.

## Section 5.3  Riemann Sums and Definite Integrals

**Objective:**    In this lesson you learned how to evaluate a definite integral using a limit.

Course Number

Instructor

Date

### I. Riemann Sums  (Pages 307–308)

Let $f$ be defined on the closed interval $[a, b]$, and let $\Delta$ be a partition of $[a, b]$ given by $a = x_0 < x_1 < x_2 < \cdots < x_{n-1} < x_n = b$, where $\Delta x_i$ is the width of the $i$th subinterval $[x_{i-1}, x_i]$. If $c_i$ is any point in the $i$th subinterval , then the sum $\sum_{i=1}^{n} f(c_i)\Delta x_i$, $x_{i-1} \leq c_i \leq x_i$, is called a _____ of $f$ for the partition $\Delta$.

*What you should learn*
How to understand the definition of a Riemann sum

The width of the largest subinterval of a partition $\Delta$ is the _____ of the partition and is denoted by _____. If every subinterval is of equal width, the partition is _____ and the norm is denoted by $\|\Delta\| = \Delta x = \dfrac{b-a}{n}$ . For a general partition, the norm is related to the number of subintervals of $[a, b]$ in the following way: _____ . So the number of subintervals in a partition approaches infinity as _____ _____ .

© 2011 Cengage Learning. All Rights Reserved. May not be scanned, copied or duplicated, or posted to a publicly accessible website, in whole or in part.

## II. Definite Integrals  (Pages 309–311)

**What you should learn**
How to evaluate a
definite integral using
limits

If $f$ is defined on the closed interval $[a, b]$ and the limit of

Riemann sums over partitions $\Delta$

$$\lim_{\|\Delta\|\to 0} \sum_{i=1}^{n} f(c_i)\Delta x_i$$

exists, then $f$ is said to be _____ and

the limit is denoted by $\lim_{\|\Delta\|\to 0} \sum_{i=1}^{n} f(c_i)\Delta x_i = \int_a^b f(x)dx$ . This limit

is called the _____. The

number $a$ is _____, and the

number $b$ is _____.

It is important to see that, although the notation is similar,

definite integrals and indefinite integrals are different concepts:

a definite integral is _____, whereas an

indefinite integral is _____.

If a function $f$ is continuous on the closed interval $[a, b]$, then $f$ is

_____ on $[a, b]$.

**Example 1:**   Evaluate the definite integral $\int_{-1}^{3} (2 - x)\, dx$ .

If $f$ is continuous and nonnegative on the closed interval $[a, b]$,
then the area of the region bounded by the graph of $f$, the $x$-axis,
and the vertical lines $x = a$ and $x = b$ is given by

Area $= \int$ _____

© 2011 Cengage Learning. All Rights Reserved. May not be scanned, copied or duplicated, or posted to a publicly accessible website, in whole or in part.

**III.  Properties of Definite Integrals**  (Pages 312–314)

*What you should learn*
How to evaluate a definite integral using properties of definite integrals

If $f$ is defined at $x = a$, then we define $\displaystyle\int_a^a f(x)\,dx = $ _____ .

If $f$ is integrable on $[a, b]$, then we define $\displaystyle\int_b^a f(x)\,dx = $ _____ .

If $f$ is integrable on the three closed intervals determined by $a$, $b$, and $c$, then

$$\int_a^b f(x)\,dx = \underline{\hspace{6cm}}.$$

If $f$ and $g$ are integrable on $[a, b]$ and $k$ is a constant, then the function $kf$ is integrable on $[a, b]$ and $\displaystyle\int_a^b kf(x)\,dx = $ _____ .

If $f$ and $g$ are integrable on $[a, b]$, then the function $f \pm g$ is integrable on $[a, b]$ and $\displaystyle\int_a^b [f(x) \pm g(x)]\,dx = $ _____ .

If $f$ and $g$ are continuous on the closed interval $[a, b]$ and $0 \le f(x) \le g(x)$ for $a \le x \le b$, the area of the region bounded by the graph of $f$ and the $x$-axis (between $a$ and $b$) must be _____ . In addition, this area must be _____ the area of the region bounded by the graph of $g$ and the $x$-axis between $a$ and $b$.

© 2011 Cengage Learning. All Rights Reserved. May not be scanned, copied or duplicated, or posted to a publicly accessible website, in whole or in part.

**Additional notes**

**Homework Assignment**

Page(s)

Exercises

© 2011 Cengage Learning. All Rights Reserved. May not be scanned, copied or duplicated, or posted to a publicly accessible website, in whole or in part.

## Section 5.4   The Fundamental Theorem of Calculus

**Objective:**    In this lesson you learned how to evaluate a definite
integral using the Fundamental Theorem of Calculus.

Course Number

Instructor

Date

### I. The Fundamental Theorem of Calculus  (Pages 318–320)

Informally, the Fundamental Theorem of Calculus states that

_____

_____

_____ .

*What you should learn*
How to evaluate a
definite integral using the
Fundamental Theorem of
Calculus

The **Fundamental Theorem of Calculus** states that if $f$ is

continuous on the closed interval $[a, b]$ and $F$ is an antiderivative

of $f$ on the interval $[a, b]$, then $\int_a^b f(x)\,dx =$ _____ .

Guidelines for Using the Fundamental Theorem of Calculus

1.   Provided you can find an antiderivative of $f$, you now have

     a way to evaluate a definite integral without _____

     _____ .

2.   When applying the Fundamental Theorem, the following notation is

     convenient.   $\int_a^b f(x)\,dx = F(x)\Big]_a^b =$ _____ .

3.   When using the Fundamental Theorem of Calculus, it is not

     necessary to include a _____ .

**Example 1:**  Find $\int_{-2}^{2}(4 - x^2)\,dx$ .

**Example 2:**  Find the area of the region bounded by the $x$-axis
and the graph of $f(x) = 2 + e^x$ for $0 \le x \le 6$ .

© 2011 Cengage Learning. All Rights Reserved. May not be scanned, copied or duplicated, or posted to a publicly accessible website, in whole or in part.

## II. The Mean Value Theorem for Integrals (Page 321)

> **What you should learn**
> How to understand and use the Mean Value Theorem for Integrals

The **Mean Value Theorem for Integrals** states that if $f$ is continuous on the closed interval $[a, b]$, then there exists a number $c$ in the closed interval $[a, b]$ such that $\int_a^b f(x)\,dx =$ 

_____ .

The Mean Value Theorem for Integrals does not specify how to determine $c$. It merely guarantees _____

_____ .

## III. Average Value of a Function (Pages 322–323)

> **What you should learn**
> How to find the average value of a function over a closed interval

If $f$ is integrable on the closed interval $[a, b]$, then the **average value** of $f$ on the interval is

Average value of $f$ on $[a, b] =$ _____ $\int$

**Example 3:** Find the average value of $f(x) = 0.24x^2 + 4$ on $[0, 10]$.

## IV. The Second Fundamental Theorem of Calculus (Pages 324–326)

> **What you should learn**
> How to understand and use the Second Fundamental Theorem of Calculus

The Second Fundamental Theorem of Calculus states that if $f$ is continuous on an open interval $I$ containing $a$, then, for every $x$ in the interval, $\dfrac{d}{dx}\left[\int_a^x f(t)\,dt\right] =$ _____ .

© 2011 Cengage Learning. All Rights Reserved. May not be scanned, copied or duplicated, or posted to a publicly accessible website, in whole or in part.

## V.  Net Change Theorem  (Pages 327–328)

*What you should learn*
How to understand and
use the Net Change
Theorem

The **Net Change Theorem** states that the definite integral of the
rate of change of a quantity $F'(x)$ gives the total change, or **net
change,** in that quantity of the interval $[a, b]$.

$$\int_a^b F'(x)\,dx = \underline{\hspace{3in}}$$

**Example 4:**   Liquid flows out of a tank at a rate of $40 - 2t$
gallons per minute, where $0 \le t \le 20$. Find the
volume of liquid that flows out of the tank during
the first 5 minutes.

**Additional notes**

© 2011 Cengage Learning. All Rights Reserved. May not be scanned, copied or duplicated, or posted to a publicly accessible website, in whole or in part.

**Additional notes**

**Homework Assignment**

Page(s)

Exercises

© 2011 Cengage Learning. All Rights Reserved. May not be scanned, copied or duplicated, or posted to a publicly accessible website, in whole or in part.

## Section 5.5  Integration by Substitution

**Objective:**    In this lesson you learned how to evaluate different types of definite and indefinite integrals using a variety of methods.

Course Number

Instructor

Date

**I.  Pattern Recognition**  (Pages 333–335)

The role of substitution in integration is comparable to the role of _____ in differentiation.

*What you should learn*
How to use pattern recognition to find an indefinite integral

**Antidifferentiation of a Composite Function**

Let $g$ be a function whose range is an interval $I$, and let $f$ be a function that is continuous on $I$. If $g$ is differentiable on its domain and $F$ is an antiderivative of $f$ on $I$, then

$$\int f(g(x))g'(x)\,dx = \text{_____}. \text{ Letting}$$

$u = g(x)$ gives $du = g'(x)\,dx$ and $\int f(u)\,du = $ _____.

**Example 1:**  Find $\int (2-3x^2)^3(-6x)\,dx$.

Many integrands contain the variable part of $g'(x)$ but are missing a constant multiple. In such cases, you can _____

_____.

**Example 2:**  Find $\int 6x^2(4x^3-1)^2\,dx$.

© 2011 Cengage Learning. All Rights Reserved. May not be scanned, copied or duplicated, or posted to a publicly accessible website, in whole or in part.

**II.  Change of Variables**  (Pages 336–337)

*What you should learn*
How to use a change of variables to find an indefinite integral

With a formal **change of variables,** you completely _____

_____

_____. The change of variables technique

uses the _____ notation for the differential. That is, if

$u = g(x)$ , then $du = $ _____, and the integral

takes the form  $\int f(g(x))g'(x)\,dx = \int$ _____.

**Example 3:**   Find  $\int 6x^2(4x^3 - 1)^2\,dx$  using change of variables.

Complete the list of guidelines for making a change of variables.

1.

2.

3.

4.

5.

6.

**III.  The General Power Rule for Integration**  (Page 338)

*What you should learn*
How to use the General Power Rule for Integration to find an indefinite integral

One of the most common $u$-substitutions involves _____

_____,

and is given a special name—the _____

_____. It states that if $g$ is a differentiable

function of $x$, then  $\int$ _____.

© 2011 Cengage Learning. All Rights Reserved. May not be scanned, copied or duplicated, or posted to a publicly accessible website, in whole or in part.

Equivalently, if $u = g(x)$, then $\int$ _____

**Example 4:**   Find $\int (4x^3 - x^2)(12x^2 - 2x)\,dx$.

## IV.  Change of Variables for Definite Integrals
   (Pages 339–340)

When using $u$-substitution with a definite integral, it is often

convenient to _____

_____ rather than to convert the antiderivative

back to the variable $x$ and evaluate at the original limits.

**Change of Variables for Definite Integrals**

If the function $u = g(x)$ has a continuous derivative on the

closed interval $[a, b]$ and $f$ is continuous on the range of $g$, then

$$\int_a^b f(g(x))g'(x)\,dx = \int \underline{\hspace{3cm}}.$$

**Example 5:**   Find $\int_0^4 2x(2x^2 - 3)^2\,dx$.

| | |
|---|---|
| *What you should learn* | |
| How to use a change of variables to evaluate a definite integral | |

© 2011 Cengage Learning. All Rights Reserved. May not be scanned, copied or duplicated, or posted to a publicly accessible website, in whole or in part.

## V.  Integration of Even and Odd Functions  (Page 341)

Occasionally, you can simplify the evaluation of a definite

integral over an interval that is symmetric about the *y*-axis or

about the origin by _____

_____ .

*What you should learn*
How to evaluate a
definite integral
involving an even or odd
function

Let *f* be integrable on the closed interval $[-a, a]$.

If *f* is an _____ function, then $\int_{-a}^{a} f(x)\, dx = 2 \int_{0}^{a} f(x)\, dx$ .

If *f* is an _____ function, then $\int_{-a}^{a} f(x)\, dx = 0$ .

---

**Homework Assignment**

Page(s)

Exercises

---

© 2011 Cengage Learning. All Rights Reserved. May not be scanned, copied or duplicated, or posted to a publicly accessible website, in whole or in part.

## Section 5.6  Numerical Integration

**Objective:**     In this lesson you learned how to approximate a definite
integral using the Trapezoidal Rule and Simpson's Rule.

Course Number

Instructor

Date

### I.  The Trapezoidal Rule  (Pages 347–349)

In your own words, describe how the Trapezoidal Rule
approximates the area under the graph of a continuous function $f$.

> **What you should learn**
> How to approximate a
> definite integral using the
> Trapezoidal Rule

The **Trapezoidal Rule** states that if $f$ is continuous on $[a, b]$, then

$$\int_a^b f(x)\, dx \approx \underline{\hspace{6cm}}.$$

Moreover, as $n \to \infty$, the right-hand side approaches $\int_a^b f(x)\, dx$.

The approximation of the area under a curve given by the

Trapezoidal Rule tends to become _____ as $n$

increases.

**Example 1:**   Use the Trapezoidal Rule to approximate
$\int_1^2 \dfrac{x}{3-x}\, dx$ using $n = 4$. Round your answer to
three decimal places.

### II.  Simpson's Rule  (Pages 349–350)

In your own words, describe how Simpson's Rule approximates
the area under the graph of a continuous function $f$.

> **What you should learn**
> How to approximate a
> definite integral using
> Simpson's Rule

© 2011 Cengage Learning. All Rights Reserved. May not be scanned, copied or duplicated, or posted to a publicly accessible website, in whole or in part.

For Simpson's Rule, what restriction is there on the value of $n$?

**Simpson's Rule** states that if $f$ is continuous on $[a, b]$ and $n$ is even, then

$$\int_a^b f(x)\,dx \approx \underline{\hspace{9cm}}.$$

Moreover, as $n \to \infty$, the right-hand side approaches $\int_a^b f(x)\,dx$.

**Example 2:** Use Simpson's Rule to approximate $\int_1^2 \dfrac{x}{3-x}\,dx$ using $n = 4$. Round your answer to three decimal places.

**III. Error Analysis** (Page 351)

For $\underline{\hspace{3cm}}$ Rule, the error $E$ in approximating

$$\int_a^b f(x)\,dx \text{ is given as } |E| \le \frac{(b-a)^5}{180n^4}\Big[\max\big|f^{(4)}(x)\big|\Big],\ a \le x \le b.$$

For $\underline{\hspace{3cm}}$ Rule, the error $E$ in approximating

$$\int_a^b f(x)\,dx \text{ is given as } |E| \le \frac{(b-a)^3}{12n^2}\Big[\max\big|f''(x)\big|\Big],\ a \le x \le b.$$

> *What you should learn*
> How to analyze errors in the Trapezoidal Rule and Simpson's Rule

**Homework Assignment**

Page(s)

Exercises

© 2011 Cengage Learning. All Rights Reserved. May not be scanned, copied or duplicated, or posted to a publicly accessible website, in whole or in part.

## Section 5.7  The Natural Logarithmic Function:  Integration

Course Number

Instructor

Date

**Objective:**     In this lesson you learned how to find the antiderivative
of the natural logarithmic function.

**I.  Log Rule for Integration**  (Pages 354–357)

*What you should learn*
How to use the Log Rule
for Integration to
integrate a rational
function

Let $u$ be a differentiable function of $x$.

$$\int \frac{1}{x}\, dx = \underline{\hspace{5cm}}$$

$$\int \frac{u'}{u}\, dx = \int \frac{1}{u}\, du = \underline{\hspace{5cm}}$$

**Example 1:**   Find $\int \left(1 - \frac{1}{x}\right) dx$ .

**Example 2:**   Find $\int \frac{x^2}{3 - x^3}\, dx$ .

**Example 3:**   Find $\int \frac{x^2 - 4x + 1}{x}\, dx$ .

If a rational function has a numerator of degree greater than

_____ ,

division may reveal a form to which you can apply the Log Rule.

© 2011 Cengage Learning. All Rights Reserved. May not be scanned, copied or duplicated, or posted to a publicly accessible website, in whole or in part.

**Guidelines for Integration**

1.

2.

3.

4.

## II. Integrals of Trigonometric Functions (Pages 358–359)

$\int \sin u \, du = $ _____

$\int \cos u \, du = $ _____

$\int \tan u \, du = $ _____

$\int \cot u \, du = $ _____

$\int \sec u \, du = $ _____

$\int \csc u \, du = $ _____

> **What you should learn**
> How to integrate
> trigonometric functions

**Example 4:** Find $\int \csc 5x \, dx$

> **Homework Assignment**
>
> Page(s)
>
> Exercises

© 2011 Cengage Learning. All Rights Reserved. May not be scanned, copied or duplicated, or posted to a publicly accessible website, in whole or in part.

## Section 5.8  Inverse Trigonometric Functions:  Integration

**Objective:**     In this lesson you learned how to find antiderivatives of inverse trigonometric functions.

Course Number

Instructor

Date

**I. Integrals Involving Inverse Trigonometric Functions**
   (Pages 363–364)

Let $u$ be a differentiable function of $x$, and let $a > 0$.

$$\int \frac{du}{\sqrt{a^2 - u^2}} = \underline{\hspace{4cm}}.$$

$$\int \frac{du}{a^2 + u^2} = \underline{\hspace{4cm}}.$$

$$\int \frac{du}{u\sqrt{u^2 - a^2}} = \underline{\hspace{4cm}}.$$

**Example 1:**   $\displaystyle\int \frac{6x\,dx}{4 + 9x^4}$

*What you should learn*
How to integrate functions whose antiderivatives involve inverse trigonometric functions

**II.  Completing the Square**  (Pages 364–365)

Completing the square helps when $\underline{\hspace{3cm}}$
$\underline{\hspace{6cm}}$.

*What you should learn*
How to use the method of completing the square to integrate a function

**Example 2:**   Complete the square for the polynomial:
$x^2 + 6x + 3$.

**Example 3:**   Complete the square for the polynomial:
$2x^2 + 16x$.

© 2011 Cengage Learning. All Rights Reserved. May not be scanned, copied or duplicated, or posted to a publicly accessible website, in whole or in part.

## III. Review of Basic Integration Rules (Pages 366–367)

Complete the following selected basic integration rules.

**What you should learn**
How to review the basic
integration rules
involving elementary
functions

$$\int \frac{du}{u} = \underline{\hspace{4cm}}$$

$$\int du = \underline{\hspace{4cm}}$$

$$\int \cot u \; du = \underline{\hspace{6cm}}$$

$$\int \frac{du}{a^2 + u^2} = \underline{\hspace{5cm}}$$

$$\int \cos u \; du = \underline{\hspace{4cm}}$$

$$\int \sec^2 u \; du = \underline{\hspace{5cm}}$$

$$\int \tan u \; du = \underline{\hspace{6cm}}$$

$$\int e^u \; du = \underline{\hspace{4cm}}$$

**Homework Assignment**

Page(s)

Exercises

© 2011 Cengage Learning. All Rights Reserved. May not be scanned, copied or duplicated, or posted to a publicly accessible website, in whole or in part.

## Section 5.9  Hyperbolic Functions

**Objective:**     In this lesson you learned about the properties,
                   derivatives, and antiderivatives of hyperbolic functions.

Course Number

Instructor

Date

**I. Hyperbolic Functions**  (Pages 371–373)

Complete the following definitions of the hyperbolic functions.

*What you should learn*
How to develop
properties of hyperbolic
functions

$\sinh x = $ _____.

$\cosh x = $ _____.

$\tanh x = $ _____.

$\operatorname{csch} x = $ _____.

$\operatorname{sech} x = $ _____.

$\coth x = $ _____.

Complete the following hyperbolic identities.

$\cosh^2 x - \sinh^2 x = $ _____.

$\tanh^2 x + \operatorname{sech}^2 x = $ _____.

$\coth^2 x - \operatorname{csch}^2 x = $ _____.

$\dfrac{-1 + \cosh 2x}{2} = $ _____.

$\dfrac{1 + \cosh 2x}{2} = $ _____.

$2 \sinh x \cosh x = $ _____.

$\cosh^2 x + \sinh^2 x = $ _____.

$\sinh (x + y) = $ _____.

$\sinh (x - y) = $ _____.

$\cosh (x + y) = $ _____.

$\cosh (x - y) = $ _____.

© 2011 Cengage Learning. All Rights Reserved. May not be scanned, copied or duplicated, or posted to a publicly accessible website, in whole or in part.

## II. Differentiation and Integration of Hyperbolic Functions
(Pages 373–375)

**What you should learn**
How to differentiate and integrate hyperbolic functions

Let $u$ be a differentiable function of $x$. Complete each of the following rules of differentiation and integration.

$\dfrac{d}{dx}[\sinh u] = $ _____.

$\dfrac{d}{dx}[\cosh u] = $ _____.

$\dfrac{d}{dx}[\tanh u] = $ _____.

$\dfrac{d}{dx}[\coth u] = $ _____.

$\dfrac{d}{dx}[\operatorname{sech} u] = $ _____.

$\dfrac{d}{dx}[\operatorname{csch} u] = $ _____.

$\displaystyle\int \cosh u \, du = $ _____.

$\displaystyle\int \sinh u \, du = $ _____.

$\displaystyle\int \operatorname{sech}^2 u \, du = $ _____.

$\displaystyle\int \operatorname{csch}^2 u \, du = $ _____.

$\displaystyle\int \operatorname{sech} u \tanh u \, du = $ _____.

$\displaystyle\int \operatorname{csch} u \coth u \, du = $ _____.

© 2011 Cengage Learning. All Rights Reserved. May not be scanned, copied or duplicated, or posted to a publicly accessible website, in whole or in part.

## III. Inverse Hyperbolic Functions (Pages 375–377)

State the inverse hyperbolic function given by each of the
following definitions and give the domain for each.

**What you should learn**
How to develop
properties of inverse
hyperbolic functions

$$\underline{\quad\quad Domain \quad\quad}$$

$\ln\left(x+\sqrt{x^2+1}\right) =$ _____ , _____ .

$\ln\left(x+\sqrt{x^2-1}\right) =$ _____ , _____ .

$\dfrac{1}{2}\ln\dfrac{1+x}{1-x} =$ _____ , _____ .

$\dfrac{1}{2}\ln\dfrac{x+1}{x-1} =$ _____ , _____ .

$\ln\dfrac{1+\sqrt{1-x^2}}{x} =$ _____ , _____ .

$\ln\left(\dfrac{1}{x}+\dfrac{\sqrt{1+x^2}}{|x|}\right) =$ _____ , _____ .

## IV. Differentiation and Integration of Inverse Hyperbolic Functions (Pages 377–378)

Let $u$ be a differentiable function of $x$. Complete each of the
following rules of differentiation and integration.

**What you should learn**
How to differentiate and
integrate functions
involving inverse
hyperbolic functions

$\dfrac{d}{dx}[\quad\underline{\quad\quad\quad\quad}\quad] = \dfrac{u'}{\sqrt{u^2+1}}$

$\dfrac{d}{dx}[\quad\underline{\quad\quad\quad\quad}\quad] = \dfrac{u'}{\sqrt{u^2-1}}$

$\dfrac{d}{dx}[\quad\underline{\quad\quad\quad\quad}\quad] = \dfrac{u'}{1-u^2}$

$\dfrac{d}{dx}[\quad\underline{\quad\quad\quad\quad}\quad] = \dfrac{u'}{1-u^2}$

$\dfrac{d}{dx}[\quad\underline{\quad\quad\quad\quad}\quad] = \dfrac{-u'}{u\sqrt{1-u^2}}$

© 2011 Cengage Learning. All Rights Reserved. May not be scanned, copied or duplicated, or posted to a publicly accessible website, in whole or in part.

$$\frac{d}{dx}\Big[ \underline{\hspace{3cm}} \Big] = \frac{-u'}{|u|\sqrt{1+u^2}}$$

$$\int \frac{du}{\sqrt{u^2 \pm a^2}} = \underline{\hspace{5cm}}.$$

$$\int \frac{du}{a^2 - u^2} = \underline{\hspace{6cm}}.$$

$$\int \frac{du}{u\sqrt{a^2 \pm u^2}} = \underline{\hspace{6cm}}.$$

**Homework Assignment**

Page(s)

Exercises

© 2011 Cengage Learning. All Rights Reserved. May not be scanned, copied or duplicated, or posted to a publicly accessible website, in whole or in part.

# Chapter 6　　Differential Equations

Course Number

Instructor

Date

## Section 6.1 · Slope Fields and Euler's Method

**Objective:** In this lesson you learned how to sketch a slope field of a
differential equation, and find a particular solution.

### I. General and Particular Solutions (Pages 388–389)

*What you should learn*
How to use initial
conditions to find
particular solutions of
differential equations

Recall that a **differential equation** in $x$ and $y$ is an equation that

_____.

A function $y = f(x)$ is a **solution** of a differential equation if

_____

_____. The

**general solution** of a differential equation is _____

_____. The

**order** of a differential equation is determined by _____

_____.

Geometrically, the general solution of a first-order differential

equation represents a family of curves known as _____

_____, one for each value assigned to the

arbitrary constant. Particular solutions of a differential equation

are obtained from _____ that give

the values of the dependent variable or one of its derivatives for

particular values of the independent variable.

**Example 1:**　For the differential equation $y'' - y' - 2y = 0$,
verify that $y = Ce^{2x}$ is a solution, and find the
particular solution determined by the initial
condition $y = 5$ when $x = 0$.

### II. Slope Fields (Pages 390–391)

*What you should learn*
How to use slope fields
to approximate solutions
of differential equations

Solving a differential equation analytically can be difficult or

even impossible. However, there is a _____

_____ you can use to learn a lot about the solution

of a differential equation. Consider a differential equation of the

© 2011 Cengage Learning. All Rights Reserved. May not be scanned, copied or duplicated, or posted to a publicly accessible website, in whole or in part.

form $y' = F(x, y)$ where $F(x, y)$ is some expression in $x$ and $y$. At
each point $(x, y)$ in the $xy$-plane where $F$ is defined, the
differential equation determines the _____
of the solution at that point. If you draw a short line segment
with slope $F(x, y)$ at selected points $(x, y)$ in the domain of $F$,
then these line segments form a _____, or a
direction field for the differential equation $y' = F(x, y)$. Each line
segment has _____ as the solution
curve through that point. A slope field shows _____
_____ and can be helpful in
getting a visual perspective of the directions of the solutions of a
differential equation.

A solution curve of a differential equation $y' = F(x, y)$ is simply

_____

_____.


### III. Euler's Method  (Page 392)

**Euler's Method** is _____

_____

_____.

From the given information, you know that the graph of the
solution passes through _____ and
has a slope of _____ at this point. This gives a
"starting point" for _____.
From this starting point, you can proceed in the direction
_____. Using a small step $h$,
move along the tangent line until you arrive at the point $(x_1, y_1)$
where $x_1 =$ _____ and $y_1 =$ _____ . If
you think of $(x_1, y_1)$ as a new starting point, you can repeat the
process to obtain _____ .

**What you should learn**
How to use Euler's
Method to approximate
solutions of differential
equations

**Homework Assignment**
Page(s)

Exercises

© 2011 Cengage Learning. All Rights Reserved. May not be scanned, copied or duplicated, or posted to a publicly accessible website, in whole or in part.

## Section 6.2  Differential Equations:  Growth and Decay

Course Number

Instructor

Date

**Objective:**    In this lesson you learned how to use an exponential
function to model growth and decay.

### I.  Differential Equations  (Page 397)

The separation of variables strategy is to _____

_____

_____.

*What you should learn*
How to use separation of
variables to solve a
simple differential
equation

**Example 1:**   Find the general solution of $\dfrac{dy}{dx} = \dfrac{3x^2 - 1}{2y + 5}$.

### II.  Growth and Decay Models  (Pages 398–401)

In many applications, the rate of change of a variable $y$ is

_____ to the value of $y$. If $y$ is a function of time

$t$, the proportion can be written as _____.

*What you should learn*
How to use exponential
functions to model
growth and decay in
applied problems

The **Exponential Growth and Decay Model** states that if $y$ is a

differentiable function of $t$ such that $y > 0$ and $y' = ky$, for some

constant $k$, then _____ where $C$ is the _____

_____, and $k$ is the _____.

**Exponential growth** occurs when _____ and

**exponential decay** occurs when _____.

**Example 2:**    The rate of change of $y$ is proportional to $y$. When
$t = 0$, $y = 5$. When $t = 4$, $y = 10$. What is the value
of $y$ when $t = 2$?

© 2011 Cengage Learning. All Rights Reserved. May not be scanned, copied or duplicated, or posted to a publicly accessible website, in whole or in part.

In a situation of radioactive decay, **half-life** is _____

_____

_____.


**Newton's Law of Cooling** states that _____

_____

_____

_____.

---

**Homework Assignment**

Page(s)

Exercises

© 2011 Cengage Learning. All Rights Reserved. May not be scanned, copied or duplicated, or posted to a publicly accessible website, in whole or in part.

## Section 6.3   Differential Equations:  Separation of Variables

Course Number

Instructor

Date

**Objective:**    In this lesson you learned how to use separation of variables to solve a differential equation.

### I.  Separation of Variables  (Pages 405–406)

*What you should learn*
How to recognize and solve differential equations that can be solved by separation of variables

Consider a differential equation that can be written in the form

$M(x) + N(y)\dfrac{dy}{dx} = 0$, where $M$ is a continuous function of $x$

alone and $N$ is a continuous function of $y$ alone. Such equations
are said to be _____, and the solution
procedure is called _____.
For this type of equation, all $x$ terms can be _____
_____, all $y$ terms can be _____
_____, and a solution can be obtained by integration.

Give an example of a separable differential equation.

**Example 1:**   Solve the differential equation $2yy' = e^x$ subject to
the initial condition $y = 3$ when $x = 0$.

### II.  Homogeneous Differential Equations  (Pages 407–408)

*What you should learn*
How to recognize and solve homogeneous differential equations

Some differential equations that are not separable in $x$ and $y$ can

be made separable by _____. This

is true for differential equations of the form $y' = f(x, y)$ where $f$

is a _____. The

function given by $f(x, y)$ is **homogeneous of degree $n$** if

_____, where $n$ is an integer.

A **homogeneous differential equation** is an equation of the

form _____, where $M$ and

$N$ are homogenous functions of the same degree.

© 2011 Cengage Learning. All Rights Reserved. May not be scanned, copied or duplicated, or posted to a publicly accessible website, in whole or in part.

**Example 2:**   State whether the function
$f(x, y) = 6xy^3 + 4x^4 - x^2y^2$ is homogeneous. If
so, what is its degree?

If $M(x, y)dx + N(x, y)dy = 0$ is homogeneous, then it can be

transformed into a differential equation whose variables are

separable by the substitution _____, where $v$ is

a differentiable function of $x$.

## III.  Applications  (Pages 409–414)

**Example 3:**   A new legal requirement is being publicized
through a public awareness campaign to a
population of 1 million citizens. The rate at which
the population hears about the requirement is
assumed to be proportional to the number of
people who are not yet aware of the requirement.
By the end of 1 year, half of the population has
heard of the requirement. How many will have
heard of it by the end of 2 years?

> **What you should learn**
> How to use differential
> equations to model and
> solve applied problems

A common problem in electrostatics, thermodynamics, and

hydrodynamics involves finding a family of curves, each of

which is _____ to all members of a given

family of curves. If one family of curves intersects another

family of curves at right angles, then the two families are said to

be _____, and each curve in

one of the families is called an _____ of

the other family.

> **Homework Assignment**
>
> Page(s)
>
> Exercises

© 2011 Cengage Learning. All Rights Reserved. May not be scanned, copied or duplicated, or posted to a publicly accessible website, in whole or in part.

Course Number

Instructor

Date

## Section 6.4  The Logistic Equation

**Objective:**    In this lesson you learned how to solve and analyze
logistic differential equations.

*What you should learn*
How to solve and analyze
logistic differential
equations

**I. Logistic Differential Equation** (Pages 419–423)

Exponential growth is unlimited, but when describing a
population, there often exists some upper limit $L$ past which
growth cannot occur. This upper limit $L$ is called the _____
_____, which is the maximum population $y(t)$ that
can be sustained or supported as time $t$ increases. A model that is
often used for this type of growth is the _____

_____ $\dfrac{dy}{dt} = ky\left(1 - \dfrac{y}{L}\right)$, where $k$ and $L$

are positive constants. A population that satisfies this equation
does not grow without bound, but approaches _____

_____ as $t$ increases.

If $y$ is between 0 and the carrying capacity $L$, then the population
_____. If $y$ is greater than $L$, then the
population _____.

The general solution of the logistic differential equation is of the

form $y = $ ————————— .

For any logistic growth curve for which the solution starts below
the carrying capacity $L$, the point of inflection occurs at

_____.

Give an example of a real-life application for which logistic
differential equations are an appropriate model.

© 2011 Cengage Learning. All Rights Reserved. May not be scanned, copied or duplicated, or posted to a publicly accessible website, in whole or in part.

**Additional notes**

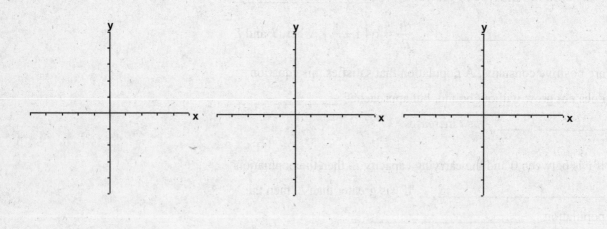

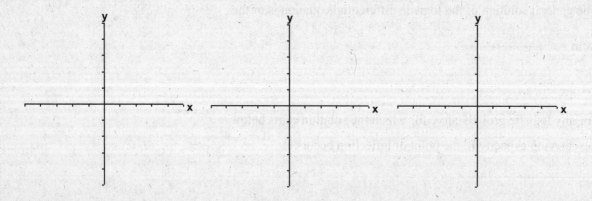

**Homework Assignment**

Page(s)

Exercises

© 2011 Cengage Learning. All Rights Reserved. May not be scanned, copied or duplicated, or posted to a publicly accessible website, in whole or in part.

## Section 6.5  First-Order Linear Differential Equations

Course Number

Instructor

Date

**Objective:**    In this lesson you learned how to solve a first-order linear differential equation and a Bernoulli differential equation.

### I.  First-Order Linear Differential Equations
(Pages 426–428)

*What you should learn*
How to solve a first-order linear differential equation

A **first-order linear differential equation** is an equation of the form _____, where $P$ and $Q$ are continuous functions of $x$. An equation that is written in this form is said to be in _____.

To solve a linear differential equation, _____

_____.

Then integrate $P(x)$ and form the expression $u(x) = e^{\int P(x)\,dx}$, which is called a(n) _____. The

general solution of the equation is $y = $ _____.

**Example 1:**    Write $e^x y' = 5 - (2 + e^x)y$ in standard form.

**Example 2:**    Find the general solution of $y' - 3y = e^{6x}$.

### II.  Bernoulli Equation  (Pages 428–429)

*What you should learn*
How to solve a Bernoulli differential equation

A well-known nonlinear equation, $y' + P(x)y = Q(x)y^n$, that reduces to a linear one with an appropriate substitution is

_____.

© 2011 Cengage Learning. All Rights Reserved. May not be scanned, copied or duplicated, or posted to a publicly accessible website, in whole or in part.

State the general solution of the Bernoulli equation.

### III. Applications  (Pages 430–432)

Give examples of types of problems that can be described in
terms of a first-order linear differential equation.

| *What you should learn* |
| --- |
| How to use linear differential equations to solve applied problems |

**Additional notes**

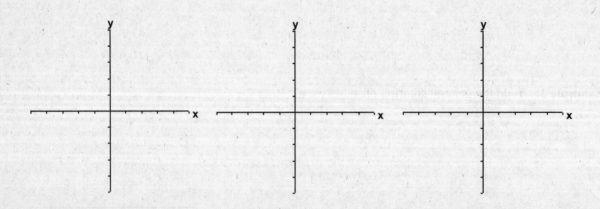

| **Homework Assignment** |
| --- |
| Page(s) |
| Exercises |

© 2011 Cengage Learning. All Rights Reserved. May not be scanned, copied or duplicated, or posted to a publicly accessible website, in whole or in part.

## Section 6.6  Predator-Prey Differential Equations

**Objective:**     In this lesson you learned how to analyze predator-prey
and competing-species differential equations.

Course Number

Instructor

Date

### I.  Predator-Prey Differential Equations  (Pages 435–437)

**What you should learn**
How to analyze predator-
prey differential equation

Consider a predator-prey relationship involving foxes (predators)
and rabbits (prey). Assume that the rabbits are the primary food
source for the foxes, the rabbits have an unlimited food supply,
and there is no threat to the rabbits other than from the foxes. Let
$x$ represent _____, let $y$ represent
_____, and let $t$ represent _____.

If there are no foxes, then the rabbit population will grow
according to _____
_____. If there are foxes but no rabbits,
the foxes have no food and their population will decay according
to _____.
If both foxes and rabbits are present, there is an interaction rate
of _____ for the rabbit population given by
_____, and an interaction rate of _____
in the fox population given by _____, where $b, n > 0$.
So, the rates of change in each population can be modeled by the
following predator-prey system of differential equations.

Rate of change of prey:     $\dfrac{dx}{dt} =$ _____

Rate of change of predators:     $\dfrac{dy}{dt} =$ _____

These equations are called predator-prey equations or _____
_____. The equations are _____
because the rates of change do not depend explicitly on time $t$.

© 2011 Cengage Learning. All Rights Reserved. May not be scanned, copied or duplicated, or posted to a publicly accessible website, in whole or in part.

Describe how to solve the predator-prey equations.

The **critical points** or **equilibrium points** of the predator-prey

equations are _____

_____.

## II. Competing Species (Pages 438–439)

Consider two species that compete with each other for the food
available in their common environment. Assume that their
populations are given by $x$ and $y$ at time $t$. If the species interact,
then their competition for resources causes a _____
_____ in each population proportional to the product
$xy$. The **competing-species equations** are as follows.

*What you should learn*
How to analyze
competing-species
differential equations

Rate of change of first species with interaction:

$\dfrac{dx}{dt} =$ _____

Rate of change of second species with interaction:

$\dfrac{dy}{dt} =$ _____

**Homework Assignment**

Page(s)

Exercises

© 2011 Cengage Learning. All Rights Reserved. May not be scanned, copied or duplicated, or posted to a publicly accessible website, in whole or in part.

# Chapter 7    Applications of Integration

Course Number

Instructor

Date

## Section 7.1  Area of a Region Between Two Curves

**Objective:** In this lesson you learned how to use a definite integral to find the area of a region bounded by two curves.

### I.  Area of a Region Between Two Curves  (Pages 448–449)

If $f$ and $g$ are continuous on $[a, b]$ and $g(x) \le f(x)$ for all $x$ in $[a, b]$, then the area of the region bounded by the graphs of $f$ and $g$ and the vertical lines $x = a$ and $x = b$ is

$$A = \int_a^b \underline{\hspace{5cm}}.$$

*What you should learn*
How to find the area of a region between two curves using integration

**Example 1:**  Find the area of the region bounded by the graphs of $y = 6 + 3x - x^2$, $y = 2x - 9$, $x = -2$, and $x = 2$.

### II.  Area of a Region Between Intersecting Curves
(Pages 450–452)

A more common problem involves the area of a region bounded by two intersecting graphs, where the values of $a$ and $b$ must be

$\underline{\hspace{6cm}}.$

*What you should learn*
How to find the area of a region between intersecting curves using integration

**Example 2:**  Find the area of the region bounded by the graphs of $y = x^2 - 5$ and $y = 1 - x$.

If two curves intersect at more than two points. Then to find the area of the region between the graphs, you must $\underline{\hspace{4cm}}$

$\underline{\hspace{10cm}}$

$\underline{\hspace{10cm}}.$

© 2011 Cengage Learning. All Rights Reserved. May not be scanned, copied or duplicated, or posted to a publicly accessible website, in whole or in part.

**III.  Integration as an Accumulation Process**  (Page 453)

*What you should learn*
How to describe
integration as an
accumulation process

In this section, the integration formula for the area between two
curves was developed by using a _____ as
the representative element. For each new application in the
remaining sections of this chapter, an appropriate representative
element will be constructed using _____
_____. Each integration formula will then
be obtained by _____ these
representative elements.

**Homework Assignment**

Page(s)

Exercises

© 2011 Cengage Learning. All Rights Reserved. May not be scanned, copied or duplicated, or posted to a publicly accessible website, in whole or in part.

Course Number

Instructor

Date

## Section 7.2  Volume:  The Disk Method

**Objective:**    In this lesson you learned how to find the volume of a solid of revolution by the disk and washer methods.

**I.  The Disk Method**  (Pages 458–460)

A **solid of revolution** is formed by _____
_____. The line is called _____
_____. The simplest such solid
is _____,
which is formed by revolving a rectangle about an axis adjacent
to one side of the rectangle.

*What you should learn*
How to find the volume of a solid of revolution using the disk method

To find the volume of a solid of revolution with the **Disk Method,** use one of the following formulas.

Horizontal axis of revolution:

Volume = _____ $\int$ _____.

Vertical axis of revolution:

Volume = _____ $\int$ _____.

The simplest application of the disk method involves a plane region bounded by _____.
If the axis of revolution is the $x$-axis, the radius $R(x)$ is simply
_____.

**Example 1:**  Find the volume of the solid formed by revolving the region bounded by the graph of
$f(x) = 0.5x^2 + 4$ and the $x$-axis, between $x = 0$ and $x = 3$, about the $x$-axis.

**II.  The Washer Method**  (Pages 461–463)

The Washer Method is used to find the volume of a solid of
revolution that has _____.

*What you should learn*
How to find the volume of a solid of revolution using the washer method

Larson/Edwards  **Calculus: Early Transcendental Functions 5e**  Notetaking Guide

© 2011 Cengage Learning. All Rights Reserved. May not be scanned, copied or duplicated, or posted to a publicly accessible website, in whole or in part.

Consider a region bounded by an outer radius $R(x)$ and an inner radius $r(x)$. The **Washer Method** states that if this region is revolved about its axis of revolution, the volume of the resulting solid is given by

Volume = _____ $\int$ _____.

Note that the integral involving the inner radius represents _____

_____ and is _____

_____ the integral involving the outer radius.

**Example 2:**  Find the volume of the solid formed by revolving the region bounded by the graphs of
$f(x) = -x^2 + 5x + 3$ and $g(x) = -x + 8$ about the x-axis.

**III. Solids with Known Cross Sections** (Pages 463–464)

With the disk method, you can find the volume of a solid having a circular cross section whose area is $A = \pi R^2$. This method can be generalized to solids of any shape, as long as you know _____

_____.

For cross sections of area $A(x)$ taken perpendicular to the x-axis,

Volume = $\int$ _____.

For cross sections of area $A(y)$ taken perpendicular to the y-axis,

Volume = $\int$ _____.

> **What you should learn**
> How to find the volume of a solid with a known cross section

**Homework Assignment**

Page(s)

Exercises

© 2011 Cengage Learning. All Rights Reserved. May not be scanned, copied or duplicated, or posted to a publicly accessible website, in whole or in part.

## Section 7.3  Volume:  The Shell Method

Course Number

Instructor

Date

**Objective:**    In this lesson you learned how to find the volume of a
solid of revolution by the shell method.

### I. The Shell Method  (Pages 469–471)

To find the volume of a solid of revolution with the **Shell
Method,** use one of the following formulas.

Horizontal axis of revolution:

Volume = _____ $\int$ _____ .

Vertical axis of revolution:

Volume = _____ $\int$ _____ .

**Example 1:**    Using the shell method, find the volume of the
solid formed by revolving the region bounded by
the graph of $y = 3 + 2x$ and the $x$-axis, between
$x = 1$ and $x = 4$, about the $y$-axis.

*What you should learn*
How to find the volume
of a solid of revolution
using the shell method

### II.  Comparison of Disk and Shell Methods  (Pages 471–473)

For the disk method, the representative rectangle is always

_____ to the axis of revolution.

For the shell method, the representative rectangle is always

_____ to the axis of revolution.

*What you should learn*
How to compare the uses
of the disk method and
the shell method

© 2011 Cengage Learning. All Rights Reserved. May not be scanned, copied or duplicated, or posted to a publicly accessible website, in whole or in part.

**Additional notes**

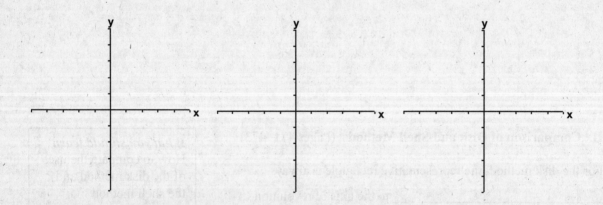

**Homework Assignment**

Page(s)

Exercises

© 2011 Cengage Learning. All Rights Reserved. May not be scanned, copied or duplicated, or posted to a publicly accessible website, in whole or in part.

## Section 7.4  Arc Length and Surfaces of Revolution

**Objective:**   In this lesson you learned how to find the length of a curve and the surface area of a surface of revolution.

Course Number

Instructor

Date

**I. Arc Length**  (Pages 478–481)

A **rectifiable** curve is _____

_____. A sufficient condition for the graph

of a function $f$ to be rectifiable between $(a, f(a))$ and $(b, f(b))$ is

that _____. Such

a function is continuously differentiable on $[a, b]$, and its graph

on the interval $[a, b]$ is a _____.

*What you should learn*
How to find the arc length of a smooth curve

Let the function given by $y = f(x)$ represent a smooth curve on the interval $[a, b]$. The arc length of $f$ between $a$ and $b$ is

$$s = \int \sqrt{\rule{3cm}{0pt}}$$

Similarly, for a smooth curve given by $x = g(y)$, the arc length of $g$ between $c$ and $d$ is

$$s = \int \sqrt{\rule{3cm}{0pt}}$$

**Example 1:**  Find the arc length of the graph of
$y = 2x^3 - x^2 + 5x - 1$ on the interval $[0, 4]$.

**II. Area of a Surface of Revolution**  (Pages 482–484)

If the graph of a continuous function is revolved about a line, the

resulting surface is a _____.

*What you should learn*
How to find the area of a surface of revolution

Let $y = f(x)$ have a continuous derivative on the interval $[a, b]$. The area $S$ of the surface of revolution formed by revolving the graph of $f$ about a horizontal or vertical axis is

© 2011 Cengage Learning. All Rights Reserved. May not be scanned, copied or duplicated, or posted to a publicly accessible website, in whole or in part.

$$s = \int \sqrt{\phantom{xxxxxxxxxx}}$$

where $r(x)$ is the distance between the graph of $f$ and the axis of revolution. If $x = g(y)$ on the interval $[c, d]$, then the surface area is

$$s = \int \sqrt{\phantom{xxxxxxxxxx}}$$

where $r(y)$ is the distance between the graph of $g$ and the axis of revolution.

**Example 2:**  Find the area of the surface formed by revolving the graph of $f(x) = 2x^2$ on the interval $[2, 4]$ about the $x$-axis.

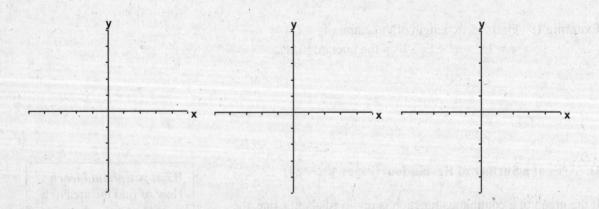

**Homework Assignment**

Page(s)

Exercises

© 2011 Cengage Learning. All Rights Reserved. May not be scanned, copied or duplicated, or posted to a publicly accessible website, in whole or in part.

Course Number

Instructor

Date

## Section 7.5  Work

**Objective:**    In this lesson you learned how to find the work done by
a constant force and by a variable force.

### I. Work Done by a Constant Force  (Page 489)

*What you should learn*
How to find the work
done by a constant force

**Work** is done by a force when _____

_____. If an object is moved a distance $D$

in the direction of an applied constant force $F$, then the work $W$

done by the force is defined as _____.

Give two examples of forces.

A **force** can be thought of as _____; a
force changes the _____ of
a body.

In the U.S. measurement system, work is typically expressed in

_____.

In the centimeter-gram-second (C-G-S) system, the basic unit of

force is the _____—the force required to produce

an acceleration of 1 centimeter per second per second on a mass

of 1 gram. In this system, work is typically expressed in _____

_____ or _____.

**Example 1:**    Find the work done in lifting a 100-pound barrel
10 feet in the air.

### II. Work Done by a Variable Force  (Pages 490–494)

*What you should learn*
How to find the work
done by a variable force

If a variable force is applied to an object, calculus is needed to

determine the work done, because _____

_____.

© 2011 Cengage Learning. All Rights Reserved. May not be scanned, copied or duplicated, or posted to a publicly accessible website, in whole or in part.

**Definition of Work Done by a Variable Force**
If an object is moved along a straight line by a continuously
varying force $F(x)$, then the **work** $W$ done by the force as the
object is moved from $x = a$ to $x = b$ is

$$W = \lim_{\|\Delta\|\to 0} \sum_{i=1}^{n} \Delta W_i$$

$$= \int \underline{\hspace{4cm}}$$

**Hooke's Law** states that the force $F$ required to compress or
stretch a spring (within its elastic limits) is proportional to the
distance $d$ that the spring is compressed or stretched from its
original length. That is, $\underline{\hspace{4cm}}$ where the
constant of proportionality $k$ (the spring constant) depends on the
specific nature of the spring.

**Newton's Law of Universal Gravitation** states that the force $F$
of attraction between two particles of masses $m_1$ and $m_2$ is
proportional to the product of the masses and inversely
proportional to the square of the distance $d$ between the two
particles. That is, $\underline{\hspace{5cm}}$.

**Coulomb's Law** states that the force between two charges $q_1$
and $q_2$ in a vacuum is proportional to the product of the charges
and inversely proportional to the square of the distance $d$
between the two charges. That is, $\underline{\hspace{4cm}}$.

**Homework Assignment**

Page(s)

Exercises

© 2011 Cengage Learning. All Rights Reserved. May not be scanned, copied or duplicated, or posted to a publicly accessible website, in whole or in part.

## Section 7.6  Moments, Centers of Mass, and Centroids

Course Number

Instructor

Date

**Objective:**    In this lesson you learned how to find centers of mass and centroids.

### I.  Mass  (Page 498)

Mass is _____

_____

_____.

Force and mass are related by the equations _____

_____.

***What you should learn***
How to understand the
definition of mass

### II.  Center of Mass in a One-Dimensional System
    (Pages 499–500)

Consider an idealized situation in which a mass $m$ is

concentrated at a point. If $x$ is the distance between this point

mass and another point $P$, the **moment of $m$ about the point $P$**

is _____ and $x$ is the length of the

_____.

***What you should learn***
How to find the center of
mass in a one-
dimensional system

Now imagine a coordinate line on which the origin corresponds

to the fulcrum. Suppose several point masses are located on the

$x$-axis. The measure of the tendency of this system to rotate

about the origin is the _____,

and it is defined as _____ .

That is $M_0 =$ _____ . If $M_0$ is 0,

the system is said to be _____.

For a system that is not in equilibrium, the **center of mass** is

defined as _____

_____.

Let the point masses $m_1, m_2, \ldots, m_n$ be located at $x_1, x_2, \ldots, x_n$.

The center of mass is $\bar{x} =$ _____ , where $m =$

_____ is the total mass of the system.

© 2011 Cengage Learning. All Rights Reserved. May not be scanned, copied or duplicated, or posted to a publicly accessible website, in whole or in part.

## III. Center of Mass in a Two-Dimensional System
　　(Page 501)

*What you should learn*
How to find the center of
mass in a two-
dimensional system

Let the point masses $m_1, m_2, \ldots, m_n$ be located at $(x_1, y_1)$,

$(x_2, y_2), \ldots, (x_n, y_n)$. The **moment about the y-axis** is $M_y =$

_____ . The **moment about**

**the x-axis** is $M_x =$ _____ .

The **center of mass** $(\bar{x}, \bar{y})$, or **center of gravity,** is

$\bar{x} =$ _____ , and $\bar{y} =$ _____ ,

where $m =$ _____ is the **total mass** of
the system.

## IV. Center of Mass of a Planar Lamina　(Pages 502–504)

*What you should learn*
How to find the center of
mass of a planar lamina

A **planar lamina** is _____

_____ . **Density** is _____

_____ ; however, for

planar laminas, density is considered to be _____

_____ . Density is denoted by _____ .

Let $f$ and $g$ be continuous functions such that $f(x) \geq g(x)$ on
$[a, b]$, and consider the planar lamina of uniform density $\rho$
bounded by the graphs of $y = f(x)$, $y = g(x)$, and $a \leq x \leq b$.

The moment about the $x$-axis is given by

$$M_x = \int \left[ \phantom{xxxx} \right] \left[ \phantom{xxxx} \right] dx$$

The moment about the $y$-axis is given by

$$M_y = \int$$

The **center of mass** $(\bar{x}, \bar{y})$ is given by $\bar{x} =$ _____ ,

and $\bar{y} =$ _____ , where $m = \int$ _____ .

© 2011 Cengage Learning. All Rights Reserved. May not be scanned, copied or duplicated, or posted to a publicly accessible website, in whole or in part.

**V. Theorem of Pappus**  (Page 505)

State the Theorem of Pappus.

*What you should learn*
How to use the Theorem of Pappus to find the volume of a solid of revolution

The Theorem of Pappus can be used to find the volume of a

torus, which is _____

_____

_____ .

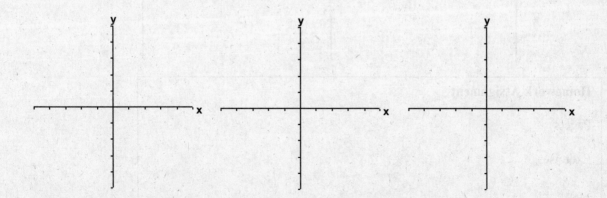

© 2011 Cengage Learning. All Rights Reserved. May not be scanned, copied or duplicated, or posted to a publicly accessible website, in whole or in part.

**Additional notes**

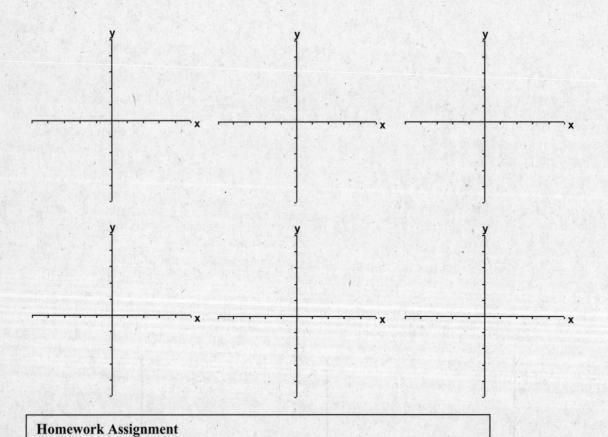

---

**Homework Assignment**

Page(s)

Exercises

© 2011 Cengage Learning. All Rights Reserved. May not be scanned, copied or duplicated, or posted to a publicly accessible website, in whole or in part.

## Section 7.7  Fluid Pressure and Fluid Force

**Objective:**    In this lesson you learned how to find fluid pressure and fluid force.

Course Number

Instructor

Date

**I. Fluid Pressure and Fluid Force**  (Pages 509–512)

**Pressure** is defined as _____

_____. The fluid

pressure on an object at a depth $h$ in a liquid is _____

_____, where $w$ is the weight-density of the liquid per

unit of volume.

*What you should learn*
How to find fluid
pressure and fluid force

When calculating fluid pressure, you can use an important

physical law called **Pascal's Principle,** which states that _____

_____

_____.

The fluid force on a submerged *horizontal* surface of area $A$ is

Fluid force $= F =$ _____.

**Example 1:**  Find the fluid force on a horizontal metal disk of diameter 3 feet that is submerged in 12 feet of seawater ($w = 64.0$).

The **force $F$ exerted by a fluid** of constant weight-density $w$ (per unit of volume) against a submerged vertical plane region from $y = c$ to $y = d$ is

$$F = w \lim_{\|\Delta\| \to 0} \sum_{i=1}^{n} h(y_i)L(y_i)\Delta y$$

$$= \underline{\phantom{w}} \int \underline{\phantom{h(y)L(y)\,dy}}$$

where $h(y)$ is the depth of the fluid at $y$ an $L(y)$ is the horizontal length of the region at $y$.

© 2011 Cengage Learning. All Rights Reserved. May not be scanned, copied or duplicated, or posted to a publicly accessible website, in whole or in part.

**Additional notes**

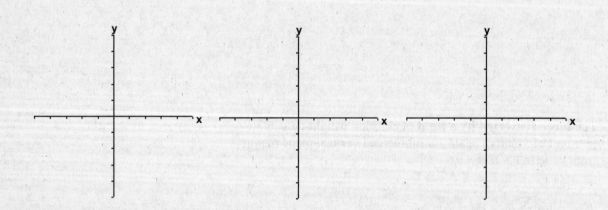

| **Homework Assignment** |
| Page(s) |
| Exercises |

© 2011 Cengage Learning. All Rights Reserved. May not be scanned, copied or duplicated, or posted to a publicly accessible website, in whole or in part.

# Chapter 8  Integration Techniques, L'Hôpital's Rule, and Improper Integrals

| |
|---|
| Course Number |
| Instructor |
| Date |

## Section 8.1  Basic Integration Rules

**Objective:** In this lesson you learned how to fit an integrand to one of the basic integration rules.

### I. Fitting Integrands to Basic Rules (Pages 520–523)

In this chapter, you study several integration techniques that greatly expand the set of integrals to which the basic integration rules can be applied. A major step in solving any integration problem is _____
_____ .

> **What you should learn**
> How to apply procedures for fitting an integrand to one of the basic integration rules

### Basic Integration Rules

$$\int kf(u)\,du = \underline{\hspace{4cm}}$$

$$\int \left[ f(u) \pm g(u) \right] du = \underline{\hspace{5cm}}$$

$$\int du = \underline{\hspace{4cm}}$$

$$\underline{\hspace{4cm}} = \frac{u^{n+1}}{n+1} + C, \quad n \neq -1$$

$$\int \frac{du}{u} = \underline{\hspace{3.5cm}}$$

$$\int e^u\,du = \underline{\hspace{3cm}}$$

$$\int a^u\,du = \underline{\hspace{3.5cm}}$$

$$\int \sin u\,du = \underline{\hspace{3.5cm}}$$

$$\int \cos u\,du = \underline{\hspace{3.5cm}}$$

© 2011 Cengage Learning. All Rights Reserved. May not be scanned, copied or duplicated, or posted to a publicly accessible website, in whole or in part.

$$\int \tan u \, du = \underline{\hspace{5cm}}$$

$$\int \cot u \, du = \underline{\hspace{5cm}}$$

$$\int \sec u \, du = \underline{\hspace{5cm}}$$

$$\int \csc u \, du = \underline{\hspace{5cm}}$$

$$\int \sec^2 u \, du = \underline{\hspace{4cm}}$$

$$\int \csc^2 u \, du = \underline{\hspace{4cm}}$$

$$\int \sec u \tan u \, du = \underline{\hspace{4cm}}$$

$$\int \csc u \cot u \, du = \underline{\hspace{4cm}}$$

$$\int \frac{du}{\sqrt{a^2 - u^2}} = \underline{\hspace{5cm}}$$

$$\int \frac{du}{a^2 + u^2} = \underline{\hspace{4cm}}$$

$$\int \frac{du}{u\sqrt{u^2 - a^2}} = \underline{\hspace{5cm}}$$

Name seven procedures for fitting integrands to basic rules. Give
an example of each procedure.

---

**Homework Assignment**

Page(s)

Exercises

---

© 2011 Cengage Learning. All Rights Reserved. May not be scanned, copied or duplicated, or posted to a publicly accessible website, in whole or in part.

## Section 8.2  Integration by Parts

**Objective:**    In this lesson you learned how to find an antiderivative using integration by parts.

Course Number

Instructor

Date

**I. Integration by Parts** (Pages 527–532)

The integration technique of **integration by parts** is particularly useful for _____

_____.

*What you should learn*
How to find an antiderivative using integration by parts

If $u$ and $v$ are functions of $x$ and have continuous derivatives, then the technique of integration by parts states that

$$\int u\, dv = \underline{\hspace{3cm}}.$$

List two guidelines for integration by parts:

1.

2.

**Example 1:**    For the indefinite integral $\int x^2 e^{2x}\, dx$, explain which factor you would choose to be $dv$ and which you would choose as $u$.

**Summary of Common Uses of Integration by Parts**
List the choices for $u$ and $dv$ in these common integration situations.

1. $\int x^n e^{ax}\, dx$,    $\int x^n \sin ax\, dx$,    or    $\int x^n \cos ax\, dx$

_____.

© 2011 Cengage Learning. All Rights Reserved. May not be scanned, copied or duplicated, or posted to a publicly accessible website, in whole or in part.

2. $\int x^n \ln x \, dx$,     $\int x^n \arcsin ax \, dx$,     or     $\int x^n \arctan ax \, dx$

_____ .

3. $\int e^{ax} \sin bx \, dx$     or     $\int e^{ax} \cos bx \, dx$

_____ .

## II.  Tabular Method  (Page 532)

In problems involving repeated applications of integration by parts, a tabular method can help organize the work. This method works well for integrals of the form $\int$ _____,

$\int$ _____ , and $\int$ _____ .

> **What you should learn**
> How to use a tabular method to perform integration by parts

---

**Homework Assignment**

Page(s)

Exercises

---

© 2011 Cengage Learning. All Rights Reserved. May not be scanned, copied or duplicated, or posted to a publicly accessible website, in whole or in part.

## Section 8.3  Trigonometric Integrals

**Objective:**    In this lesson you learned how to evaluate trigonometric integrals.

Course Number

Instructor

Date

### I.  Integrals Involving Powers of Sine and Cosine
(Pages 536–538)

In this section you studied techniques for evaluating integrals of

the form $\int \sin^m x \cos^n x \, dx$  and  $\int \sec^m x \tan^n x \, dx$  where either

$m$ or $n$ is a positive integer. To find antiderivatives for these

forms, _____

_____

_____.

> **What you should learn**
> How to solve trigonometric integrals involving powers of sine and cosine

To break up $\int \sin^m x \cos^n x \, dx$ into forms to which you can apply

the Power Rule, use the following identities.

$\sin^2 x + \cos^2 x = $ _____

$\sin^2 x = $ _____

$\cos^2 x = $ _____

List three guidelines for evaluating integrals involving sine and cosine.

© 2011 Cengage Learning. All Rights Reserved. May not be scanned, copied or duplicated, or posted to a publicly accessible website, in whole or in part.

**Wallis's Formulas** state that if $n$ is odd ($n \geq 3$), then

$$\int_0^{\pi/2} \cos^n x \, dx = \underline{\hspace{6cm}}$$

and that if $n$ is even ($n \geq 2$), then

$$\int_0^{\pi/2} \cos^n x \, dx = \underline{\hspace{6cm}}$$

## II. Integrals Involving Powers of Secant and Tangent
(Pages 539–541)

List five guidelines for evaluating integrals involving secant and tangent of the form $\int \sec^m x \tan^n x \, dx$.

> **What you should learn**
> How to solve trigonometric integrals involving powers of secant and tangent

For integrals involving powers of cotangents and cosecants,

_____

_____.

Another strategy that can be useful when integrating

trigonometric functions is _____

_____.

© 2011 Cengage Learning. All Rights Reserved. May not be scanned, copied or duplicated, or posted to a publicly accessible website, in whole or in part.

### III.  Integrals Involving Sine-Cosine Products with Different Angles  (Page 541)

Complete each of the following product-to-sum identities.

$\sin mx \, \sin nx = $ _____

$\sin mx \, \cos nx = $ _____

$\cos mx \, \cos nx = $ _____

**What you should learn**
How to solve trigonometric integrals involving sine-cosine products with different angles

© 2011 Cengage Learning. All Rights Reserved. May not be scanned, copied or duplicated, or posted to a publicly accessible website, in whole or in part.

**Additional notes**

| Homework Assignment |
| --- |
| Page(s) |
| Exercises |

© 2011 Cengage Learning. All Rights Reserved. May not be scanned, copied or duplicated, or posted to a publicly accessible website, in whole or in part.

## Section 8.4  Trigonometric Substitution

**Objective:**    In this lesson you learned how to use trigonometric
substitution to evaluation an integral.

Course Number

Instructor

Date

### I. Trigonometric Substitution  (Pages 545–549)

Now that you can evaluate integrals involving powers of

trigonometric functions, you can use trigonometric substitution

to evaluate integrals involving the radicals $\sqrt{a^2 - u^2}$ , $\sqrt{a^2 + u^2}$ ,

and $\sqrt{u^2 - a^2}$ . The objective with trigonometric substitution is

_____ .

You do this with the _____

_____ .

*What you should learn*
How to use trigonometric
substitution to solve an
integral

**Trigonometric substitution ($a > 0$):**

1.  For integrals involving $\sqrt{a^2 - u^2}$ , let $u =$ _____ .

    Then $\sqrt{a^2 - u^2} =$ _____ , where $-\pi/2 \le \theta \le \pi/2$ .

2.  For integrals involving $\sqrt{a^2 + u^2}$ , let $u =$ _____ .

    Then $\sqrt{a^2 + u^2} =$ _____ , where $-\pi/2 < \theta < \pi/2$ .

3.  For integrals involving $\sqrt{u^2 - a^2}$ , let $u =$ _____ .

    Then $\sqrt{u^2 - a^2} =$ _____ if $u > a$, where $0 \le \theta < \pi/2$;

    or $\sqrt{u^2 - a^2} =$ _____ if $u < -a$, where $\pi/2 < \theta \le \pi$.

**Special Integration Formulas ($a > 0$)**

$$\int \sqrt{a^2 - u^2}\, du =$$

_____

$$\int \sqrt{u^2 - a^2}\, du =$$

_____

© 2011 Cengage Learning. All Rights Reserved. May not be scanned, copied or duplicated, or posted to a publicly accessible website, in whole or in part.

$$\int \sqrt{u^2 + a^2}\, du = \underline{\hspace{5cm}}$$

## II. Applications (Page 550)

Give two examples of applications of trigonometric substitution.

***What you should learn***
How to use integrals to model and solve real-life applications

---

**Homework Assignment**

Page(s)

Exercises

© 2011 Cengage Learning. All Rights Reserved. May not be scanned, copied or duplicated, or posted to a publicly accessible website, in whole or in part.

## Section 8.5  Partial Fractions

Course Number

Instructor

Date

**Objective:**    In this lesson you learned how to use partial fraction decomposition to integrate rational functions.

**I. Partial Fractions**  (Pages 554–555)

The **method of partial fractions** is a procedure for _____

_____

_____ .

*What you should learn*
How to understand the
concept of a partial
fraction decomposition

**Decomposition of $N(x)/D(x)$ into Partial Fractions**

1.  **Divide if improper:**  If $N(x)/D(x)$ is _____

    _____ (that is, if the degree of the numerator is

    greater than or equal to the degree of the denominator),

    divide _____ to

    obtain $\dfrac{N(x)}{D(x)} =$ _____ ,

    where the degree of $N_1(x)$ is less than the degree of $D(x)$.

    Then apply steps 2, 3, and 4 to the proper rational expression

    $N_1(x)/D(x)$.

2.  **Factor denominator:**  Completely factor the denominator

    into factors of the form _____

    where $ax^2 + bx + c$ is irreducible.

3.  **Linear factors:**  For each factor of the form $(px + q)^m$, the

    partial fraction decomposition must include the following

    sum of $m$ fractions.

    _____

4.  **Quadratic factors:**  For each factor of the form

    $(ax^2 + bx + c)^n$, the partial fraction decomposition must

    include the following sum of $n$ fractions.

    _____

© 2011 Cengage Learning. All Rights Reserved. May not be scanned, copied or duplicated, or posted to a publicly accessible website, in whole or in part.

**II. Linear Factors**  (Pages 556–557)

To find the **basic equation** of a partial fraction

decomposition, _____

_____

_____

After finding the basic equation, _____

_____

_____

_____ .

> **What you should learn**
> How to use partial fraction decomposition with linear factors to integrate rational functions

**Example 1:**   Write the form of the partial fraction

decomposition for $\dfrac{x-4}{x^2-8x+12}$ .

**Example 2:**   Write the form of the partial fraction

decomposition for $\dfrac{2x+1}{x^3-3x^2+x-3}$ .

**Example 3:**   Solve the basic equation
$5x+3=A(x-1)+B(x+3)$ for $A$ and $B$.

**III. Quadratic Factors**  (Pages 558–560)

**Guidelines for Solving the Basic Equation**
List two guidelines for solving basic equations that involve linear factors.

> **What you should learn**
> How to use partial fraction decomposition with quadratic factors to integrate rational functions

List four guidelines for solving basic equations that involve quadratic factors.

> **Homework Assignment**
> Page(s)
> Exercises

© 2011 Cengage Learning. All Rights Reserved. May not be scanned, copied or duplicated, or posted to a publicly accessible website, in whole or in part.

## Section 8.6   Integration by Tables and Other Integration Techniques

**Objective:**   In this lesson you learned how to evaluate an indefinite integral using a table of integrals and reduction formulas.

Course Number

Instructor

Date

**I. Integration by Tables**  (Pages 563–564)

**Integration by tables** is the procedure of integrating by means of

_____ .

Integration by tables requires _____

_____ .

A computer algebra system consists, in part, of a database of

integration tables. The primary difference between using a

computer algebra system and using a table of integrals is _____

_____

_____

_____ .

*What you should learn*
How to evaluate an
indefinite integral using a
table of integrals

**Example 1:**   Use the integration table in Appendix B to identify an integration formula that could be used to find

$$\int \frac{x}{3-x}\, dx\,,$$ and identify the substitutions you would use.

**Example 2:**   Use the integration table in Appendix B to identify an integration formula that could be used to find

$$\int 3x^5 \ln x\, dx\,,$$ and identify the substitutions you would use.

© 2011 Cengage Learning. All Rights Reserved. May not be scanned, copied or duplicated, or posted to a publicly accessible website, in whole or in part.

## II. Reduction Formulas (Page 565)

An integration table formula of the form

$$\int f(x)\,dx = g(x) + \int h(x)\,dx$$, in which the right side of the

formula contains an integral, is called a _____

_____ because they _____

_____.

> **What you should learn**
> How to evaluate an indefinite integral using reduction formulas

## III. Rational Functions of Sine and Cosine (Page 566)

If you are unable to find an integral in the integration tables that involves a rational expression of sin $x$ and cos $x$, try using the following special substitution to convert the trigonometric expression to a standard rational expression.

The substitution

$u =$ _____ $=$ _____

yields

$\cos x =$ _____,

$\sin x =$ _____,

and $dx =$ _____.

> **What you should learn**
> How to evaluate an indefinite integral involving rational functions of sine and cosine

**Homework Assignment**

Page(s)

Exercises

© 2011 Cengage Learning. All Rights Reserved. May not be scanned, copied or duplicated, or posted to a publicly accessible website, in whole or in part.

## Section 8.7    Indeterminate Forms and L'Hôpital's Rule

**Objective:**    In this lesson you learned how to apply L'Hôpital's Rule
to evaluate a limit.

Course Number

Instructor

Date

### I. Indeterminate Forms  (Page 569)

The forms 0/0 and ∞/∞ are called _____ because

they _____

_____.

**What you should learn**
How to recognize limits
that produce
indeterminate forms

Occasionally an indeterminate form may be evaluated by

_____

_____. However, not all indeterminate

forms can be evaluated in this manner. This is often true when

_____ are

involved.

### II. L'Hôpital's Rule  (Pages 570–575)

The **Extended Mean Value Theorem** states that if $f$ and $g$ are

differentiable on an open interval $(a, b)$ and continuous on $[a, b]$

such that $g'(x) \neq 0$ for any $x$ in $(a, b)$, then there exists a point $c$

in $(a, b)$ such that $\dfrac{f'(c)}{g'(c)} =$ _____.

**What you should learn**
How to apply L'Hôpital's
Rule to evaluate a limit

Let $f$ and $g$ be functions that are differentiable on an open

interval $(a, b)$ containing $c$, except possibly at $c$ itself. Assume

that $g'(x) \neq 0$ for all $x$ in $(a, b)$, except possibly at $c$ itself.

**L'Hôpital's Rule** states that if the limit of $f(x)/g(x)$ as $x$

approaches $c$ produces the indeterminate form 0/0, then

$\lim\limits_{x \to c} \dfrac{f(x)}{g(x)} = \lim\limits_{x \to c}$ _____, provided the

limit on the right exists (or is infinite). This result also applies if

the limit of $f(x)/g(x)$ as $x$ approaches $c$ produces any one of the

indeterminate forms _____.

© 2011 Cengage Learning. All Rights Reserved. May not be scanned, copied or duplicated, or posted to a publicly accessible website, in whole or in part.

This theorem states that under certain conditions the limit of the

quotient $f(x)/g(x)$ is determined by _____

_____.

**Example 1:** Evaluate $\lim\limits_{x \to 0} \dfrac{1 - \cos x}{2x^2 - 3x}$.

**Example 2:** Evaluate $\lim\limits_{x \to 0} \dfrac{-3x^2}{\sqrt{x + 4} - (x/4) - 2}$.

**Homework Assignment**

Page(s)

Exercises

© 2011 Cengage Learning. All Rights Reserved. May not be scanned, copied or duplicated, or posted to a publicly accessible website, in whole or in part.

## Section 8.8   Improper Integrals

**Objective:**    In this lesson you learned how to evaluate an improper
                 integral.

Course Number

Instructor

Date

### I. Improper Integrals with Infinite Limits of Integration
(Pages 580–583)

List two properties that make an integral an **improper integral.**

1. _____

2. _____

If an integrand has an **infinite discontinuity,** then _____

_____.

*What you should learn*
How to evaluate an
improper integral that has
an infinite limit of
integration

Complete the following statements about improper integrals
having infinite limits of integration.

1.  If $f$ is continuous on the interval $[a, \infty)$, then

$$\int_a^\infty f(x)\, dx = \text{_____}$$

2.  If $f$ is continuous on the interval $(-\infty, b]$, then

$$\int_{-\infty}^b f(x)\, dx = \text{_____}$$

3.  If $f$ is continuous on the interval $(-\infty, \infty)$, then

$$\int_{-\infty}^\infty f(x)\, dx = \text{_____}$$

In the first two cases, if the limit exists, then the improper

integral _____; otherwise, the improper

integral _____. In the third case, the integral on

the left will diverge if _____

_____.

© 2011 Cengage Learning. All Rights Reserved. May not be scanned, copied or duplicated, or posted to a publicly accessible website, in whole or in part.

## II.  Improper Integrals with Infinite Discontinuities
(Pages 583–586)

Complete the following statements about improper integrals
having infinite discontinuities at or between the limits of
integration.

*What you should learn*
How to evaluate an
improper integral that has
an infinite discontinuity

1.  If $f$ is continuous on the interval $[a, b)$ and has an infinite
    discontinuity at $b$, then

    $$\int_a^b f(x)\, dx = \underline{\hspace{5cm}}$$

2.  If $f$ is continuous on the interval $(a, b]$ and has an infinite
    discontinuity at $a$, then

    $$\int_a^b f(x)\, dx = \underline{\hspace{5cm}}$$

3.  If $f$ is continuous on the interval $[a, b]$, except for some $c$
    in $(a, b)$ at which $f$ has an infinite discontinuity, then

    $$\int_a^b f(x)\, dx = \underline{\hspace{5cm}}$$

In the first two cases, if the limit exists, then the improper

integral _____; otherwise, the improper

integral _____. In the third case, the improper

integral on the left diverges if _____

_____.

Homework Assignment

Page(s)

Exercises

© 2011 Cengage Learning. All Rights Reserved. May not be scanned, copied or duplicated, or posted to a publicly accessible website, in whole or in part.

# Chapter 9    Infinite Series

## Section 9.1   Sequences

<table>
<tr><td>Course Number</td></tr>
<tr><td>Instructor</td></tr>
<tr><td>Date</td></tr>
</table>

**Objective:** In this lesson you learned how to determine whether a sequence converges or diverges.

### I. Sequences  (Page 596)

A **sequence** $\{a_n\}$ is a function whose domain is _____

_____. The numbers $a_1, a_2, a_3, \ldots,$
$a_n, \ldots$ are the _____ of the sequence. The number $a_n$
is the _____ of the sequence, and the entire
sequence is denoted by _____.

**What you should learn**
How to list the terms of a sequence

**Example 1:**   Find the first four terms of the sequence defined
by $a_n = n^2 - 4$

### II. Limit of a Sequence  (Pages 597–600)

If a sequence **converges**, its terms _____
_____.

**What you should learn**
How to determine whether a sequence converges or diverges

Let $L$ be a real number. The **limit** of a sequence $\{a_n\}$ is $L$, written
as $\lim\limits_{n\to\infty} a_n = L$ if for each $\varepsilon > 0$, there exists $M > 0$ such that

_____. If the limit $L$
of a sequence exists, then the sequence _____.
If the limit of a sequence does not exist, then the sequence

_____.

If a sequence $\{a_n\}$ agrees with a function $f$ at every positive
integer, and if $f(x)$ approaches a limit $L$ as $x\to\infty$, the sequence
must _____.

© 2011 Cengage Learning. All Rights Reserved. May not be scanned, copied or duplicated, or posted to a publicly accessible website, in whole or in part.

**Example 2:**   Find the limit of each sequence (if it exists) as $n$
approaches infinity.

    a.  $a_n = n^2 - 4$          b.  $a_n = \dfrac{2n^2}{3n - n^2}$

Complete the following properties of limits of sequences. Let
$\lim\limits_{n\to\infty} a_n = L$ and $\lim\limits_{n\to\infty} b_n = K$.

1.  $\lim\limits_{n\to\infty}(a_n \pm b_n) = $ _____

2.  $\lim\limits_{n\to\infty} c a_n = $ _____

3.  $\lim\limits_{n\to\infty}(a_n b_n) = $ _____

4.  $\lim\limits_{n\to\infty} \dfrac{a_n}{b_n} = $ _____

If $n$ is a positive integer, then **$n$ factorial** is defined as

_____. As a special

case, **zero factorial** is defined as $0! = $ _____.

Another useful limit theorem that can be rewritten for sequences
is the **Squeeze Theorem,** which states that if $\lim\limits_{n\to\infty} a_n = L = \lim\limits_{n\to\infty} b_n$
and there exists an integer $N$ such that $a_n \le c_n \le b_n$ for all $n > N$,
then $\lim\limits_{n\to\infty} c_n = $ _____.

For the sequence $\{a_n\}$, if $\lim\limits_{n\to\infty}|a_n| = 0$ then $\lim\limits_{n\to\infty} a_n = $ _____.

**III. Pattern Recognition for Sequences**  (Pages 600–601)

**Example 3:**   Determine an $n$th term for the sequence

    $0, \dfrac{1}{4}, -\dfrac{2}{9}, \dfrac{3}{16}, -\dfrac{4}{25}, \ldots$

> **What you should learn**
> How to write a formula
> for the $n$th term of
> sequence

© 2011 Cengage Learning. All Rights Reserved. May not be scanned, copied or duplicated, or posted to a publicly accessible website, in whole or in part.

## IV. Monotonic Sequences and Bounded Sequences
   (Pages 602–603)

**What you should learn**
How to use properties of
monotonic sequences and
bounded sequences

A sequence $\{a_n\}$ is **monotonic** if its terms are _____

_____ or if its terms are

_____ .

A sequence $\{a_n\}$ is _____ if there is a

real number $M$ such that $a_n \leq M$ for all $n$. The number $M$ is

called _____ of the sequence. A

sequence $\{a_n\}$ is _____ if there is a real

number $N$ such that $N \leq a_n$ for all $n$. The number $N$ is called

_____ of the sequence. A sequence $\{a_n\}$

is _____ if it is bounded above and

bounded below.

If a sequence $\{a_n\}$ is _____ ,

then it converges.

© 2011 Cengage Learning. All Rights Reserved. May not be scanned, copied or duplicated, or posted to a publicly accessible website, in whole or in part.

**Additional notes**

```
┌──────────────────────┐
│                      │
│                      │
│                      │
│                      │
└──────────────────────┘
```

```
┌─────────────────────────────────────────────────────────────┐
│                                                               │
│  Homework Assignment                                          │
│                                                               │
│  Page(s)                                                      │
│                                                               │
│  Exercises                                                    │
│                                                               │
└─────────────────────────────────────────────────────────────┘
```

© 2011 Cengage Learning. All Rights Reserved. May not be scanned, copied or duplicated, or posted to a publicly accessible website, in whole or in part.

## Section 9.2  Series and Convergence

**Objective:**    In this lesson you learned how to determine whether an infinite series converges or diverges.

Course Number

Instructor

Date

**I. Infinite Series**  (Pages 608–610)

**What you should learn**
How to understand the
definition of a convergent
infinite series

If $\{a_n\}$ is an infinite sequence, then the infinite summation

$$\sum_{n=1}^{\infty} a_n = a_1 + a_2 + a_3 + \cdots + a_n + \cdots \text{ is called an } \underline{\hspace{3cm}}$$

$\underline{\hspace{4cm}}$. The numbers $a_1$, $a_2$, $a_3$, and so on, are the

$\underline{\hspace{4cm}}$ of the series. The **sequence of partial sums**

of the series is denoted by $\underline{\hspace{6cm}}$

$\underline{\hspace{6cm}}$.

If the sequence of partial sums $\{S_n\}$ converges to $S$, then the

infinite series $\underline{\hspace{4cm}}$ to $S$. This limit is

denoted by $\lim_{n \to \infty} S_n = \sum_{n=1}^{\infty} a_n = S$ , and $S$ is called the $\underline{\hspace{3cm}}$

$\underline{\hspace{4cm}}$. If the limit of the sequence of partial sums

$\{S_n\}$ does not exist, then the series $\underline{\hspace{4cm}}$.

A **telescoping series** is of the form $(b_1 - b_2) + (b_2 - b_3) + (b_3 - b_4) + (b_4 - b_5) + \cdots$,

where $b_2$ is cancelled $\underline{\hspace{8cm}}$

$\underline{\hspace{6cm}}$. Because the $n$th partial sum of this series is

$S_n = b_1 - b_{n+1}$, it follows that a telescoping series will converge if and only if $b_n$

$\underline{\hspace{8cm}}$. Moreover, if the series

converges, its sum is $\underline{\hspace{4cm}}$.

**II. Geometric Series**  (Pages 610–612)

**What you should learn**
How to use properties of
infinite geometric series

If $a$ is a nonzero real number, then the infinite series

$$\sum_{n=0}^{\infty} ar^n = a + ar + ar^2 + \cdots + ar^n + \cdots \text{ is called a } \underline{\hspace{3cm}}$$

$\underline{\hspace{4cm}}$ with ratio $r$.

© 2011 Cengage Learning. All Rights Reserved. May not be scanned, copied or duplicated, or posted to a publicly accessible website, in whole or in part.

An infinite geometric series given by $\sum_{n=0}^{\infty} ar^n$ diverges if

_____. If _____, then the

series converges to the sum $\sum_{n=0}^{\infty} ar^n = \dfrac{a}{1-r}$ .

Given the convergent infinite series $\sum_{n=1}^{\infty} a_n = A$ and $\sum_{n=1}^{\infty} b_n = B$

and real number $c$,

$\sum_{n=1}^{\infty} ca_n = $ _____

$\sum_{n=1}^{\infty} (a_n + b_n) = $ _____

$\sum_{n=1}^{\infty} (a_n - b_n) = $ _____

**III.  *n*th-Term Test for Divergence** (Pages 612–613)

The ***n*th Term Test for Divergence** states that if $\lim_{n \to \infty} a_n \neq 0$,

then the series $\sum_{n=1}^{\infty} a_n$ _____.

> **What you should learn**
> How to use the *n*th-Term Test for Divergence of an infinite series

**Example 1:**   Determine whether the series $\sum_{n=1}^{\infty} \dfrac{2n^2}{3n^2 - 1}$

diverges.

---

**Homework Assignment**

Page(s)

Exercises

---

© 2011 Cengage Learning. All Rights Reserved. May not be scanned, copied or duplicated, or posted to a publicly accessible website, in whole or in part.

## Section 9.3   The Integral Test and *p*-Series

Course Number

Instructor

Date

**Objective:**   In this lesson you learned how to determine whether an infinite series converges or diverges.

---

**Important Vocabulary**          Define each term or concept.

**General harmonic series**

---

### I. The Integral Test  (Pages 619–620)

The **Integral Test** states that if *f* is positive, continuous and

decreasing for $x \geq 1$ and $a_n = f(n)$, then $\sum\limits_{n=1}^{\infty} a_n$ and

$\int\limits_{1}^{\infty} f(x)\, dx$  either _____.

Remember that the convergence or divergences of $\sum a_n$ is not

affected by deleting _____.
Similarly, if the conditions for the Integral Test are satisfied for

all _____, you can simply use the integral

$\int\limits_{N}^{\infty} f(x)\, dx$  to test _____.

*What you should learn*
How to use the Integral Test to determine whether an indefinite series converges or diverges

### II. *p*-Series and Harmonic Series  (Pages 621–622)

Let *p* be a positive constant. An infinite series of the form

$$\sum_{n=1}^{\infty} \frac{1}{n^p} = \frac{1}{1^p} + \frac{1}{2^p} + \frac{1}{3^p} + \cdots \text{ is called a } \underline{\hspace{3cm}}.$$

If $p = 1$, then the series $\sum\limits_{n=1}^{\infty} \dfrac{1}{n} = 1 + \dfrac{1}{2} + \dfrac{1}{3} + \cdots$ is called the

_____.

*What you should learn*
How to use properties of *p*-series and harmonic series

---

© 2011 Cengage Learning. All Rights Reserved. May not be scanned, copied or duplicated, or posted to a publicly accessible website, in whole or in part.

The **Test for Convergence of a $p$-Series** states that the $p$-series

$$\sum_{n=1}^{\infty}\frac{1}{n^p}=\frac{1}{1^p}+\frac{1}{2^p}+\frac{1}{3^p}+\frac{1}{4^p}+\cdots \text{ diverges if } \underline{\hspace{2cm}},$$

or converges if _____.

**Example 1:**   Determine whether the series $\displaystyle\sum_{n=1}^{\infty} n^{-\sqrt{2}}$ converges

or diverges.

---

**Homework Assignment**

Page(s)

Exercises

---

© 2011 Cengage Learning. All Rights Reserved. May not be scanned, copied or duplicated, or posted to a publicly accessible website, in whole or in part.

## Section 9.4   Comparison of Series

**Objective:**   In this lesson you learned how to determine whether an infinite series converges or diverges.

| Course Number |
| --- |
| Instructor |
| Date |

### I. Direct Comparison Test  (Pages 626–627)

This section presents two additional tests for positive-term series which greatly expand the variety of series you are able to test for convergence or divergence; they allow you to _____ _____ _____.

*What you should learn*
How to use the Direct Comparison Test to determine whether a series converges or diverges

Let $0 < a_n \le b_n$ for all $n$. The **Direct Comparison Test** states that

if $\displaystyle\sum_{n=1}^{\infty} b_n$ _____, then $\displaystyle\sum_{n=1}^{\infty} a_n$ _____.

If $\displaystyle\sum_{n=1}^{\infty} a_n$ _____, then $\displaystyle\sum_{n=1}^{\infty} b_n$ _____.

Use your own words to give an interpretation of this test.

### II. Limit Comparison Test  (Pages 628–629)

Suppose that $a_n > 0$ and $b_n > 0$. The **Limit Comparison Test**

states that if $\displaystyle\lim_{n\to\infty}\left(\frac{a_n}{b_n}\right) = L$, where $L$ is *finite* and *positive,* then

the two series $\displaystyle\sum a_n$ and $\displaystyle\sum b_n$ either _____ _____.

*What you should learn*
How to use the Limit Comparison Test to determine whether a series converges or diverges

© 2011 Cengage Learning. All Rights Reserved. May not be scanned, copied or duplicated, or posted to a publicly accessible website, in whole or in part.

Describe circumstances under which you might apply the Limit Comparison Test.

The Limit Comparison Test works well for comparing a "messy"

algebraic series with a $p$-series. In choosing an appropriate

$p$-series, you must choose one with _____

_____.

In other words, when choosing a series for comparison, you can

disregard all but _____

_____.

---

**Homework Assignment**

Page(s)

Exercises

---

© 2011 Cengage Learning. All Rights Reserved. May not be scanned, copied or duplicated, or posted to a publicly accessible website, in whole or in part.

## Section 9.5  Alternating Series

**Objective:**      In this lesson you learned how to determine whether an
infinite series converges or diverges.

Course Number

Instructor

Date

---

**Important Vocabulary**          Define each term or concept.

**Alternating series**

**Absolutely convergent**

**Conditionally convergent**

---

**I. Alternating Series** (Pages 633–634)

Alternating series occur in two ways: _____
_____.

Let $a_n > 0$. The **Alternating Series Test** states that the

alternating series $\sum\limits_{n=1}^{\infty} (-1)^n a_n$  and  $\sum\limits_{n=1}^{\infty} (-1)^{n+1} a_n$  converge if the

following two conditions are met:

1.

2.

**Example 1:**   Determine whether the series $\sum\limits_{n=1}^{\infty} \dfrac{(-1)^n}{n^2}$ converges

or diverges.

*What you should learn*
How to use the
Alternating Series Test to
determine whether an
infinite series converges

**II. Alternating Series Remainder** (Page 635)

For a convergent alternating series, the partial sum $S_N$ can be

_____.

*What you should learn*
How to use the
Alternating Series
Remainder to
approximate the sum of
an alternating series

© 2011 Cengage Learning. All Rights Reserved. May not be scanned, copied or duplicated, or posted to a publicly accessible website, in whole or in part.

If a convergent alternating series satisfies the condition

$a_{n+1} \leq a_n$, then the absolute value of the remainder $R_N$ involved

in approximating the sum $S$ by $S_N$ is _____

_____. That is,

$|S - S_N| = |R_N| \leq a_{N+1}$.

### III. Absolute and Conditional Convergence  (Pages 636–637)

If the series $\Sigma \, |a_n|$ converges, then the series $\Sigma \, a_n$ _____

_____.

**What you should learn**
How to classify a
convergent series as
absolutely or
conditionally convergent

**Example 2:**   Is the series $\displaystyle\sum_{n=1}^{\infty} \frac{(-1)^n}{n^2}$ absolutely or conditionally

convergent?

### IV. Rearrangement of Series  (Pages 637–638)

The terms of an infinite series can be rearranged without

changing the value of the sum of the terms only if _____

_____. If the series is

_____, then it is possible

that rearranging the terms of the series can change the value of

the sum.

**What you should learn**
How to rearrange an
infinite series to obtain a
different sum

**Homework Assignment**

Page(s)

Exercises

© 2011 Cengage Learning. All Rights Reserved. May not be scanned, copied or duplicated, or posted to a publicly accessible website, in whole or in part.

## Section 9.6  The Ratio and Root Tests

**Objective:**    In this lesson you learned how to determine whether an infinite series converges or diverges.

Course Number

Instructor

Date

**I. The Ratio Test**  (Pages 641–643)

Let $\displaystyle\sum_{n=1}^{\infty} a_n$ be an infinite series with nonzero terms. The **Ratio**

**Test** states that:

***What you should learn***
How to use the Ratio Test to determine whether a series converges or diverges

1.  The series converges absolutely if $\displaystyle\lim_{n\to\infty}\left|\dfrac{a_{n+1}}{a_n}\right|$ _____.

2.  The series diverges if $\displaystyle\lim_{n\to\infty}\left|\dfrac{a_{n+1}}{a_n}\right|$ _____ or $\displaystyle\lim_{n\to\infty}\left|\dfrac{a_{n+1}}{a_n}\right|$ _____.

3.  The test is inconclusive if $\displaystyle\lim_{n\to\infty}\left|\dfrac{a_{n+1}}{a_n}\right|$ _____.

**Example 1:**    Use the Ratio Test to determine whether the series

$\displaystyle\sum_{n=0}^{\infty}\dfrac{4^n}{n!}$  converges or diverges.

The Ratio Test is particularly useful for series that _____

_____, such as those that involve _____

_____.

**II. The Root Test**  (Page 644)

The Root Test for convergence or divergence of series works

especially well for series involving _____.

***What you should learn***
How to use the Root Test to determine whether a series converges or diverges

Let $\displaystyle\sum_{n=1}^{\infty} a_n$ be an infinite series. The **Root Test** states that:

1.  The series converges absolutely if $\displaystyle\lim_{n\to\infty}\sqrt[n]{|a_n|}$ _____.

2.  The series diverges if $\displaystyle\lim_{n\to\infty}\sqrt[n]{|a_n|}$ _____ or $\displaystyle\lim_{n\to\infty}\sqrt[n]{|a_n|}$ _____.

© 2011 Cengage Learning. All Rights Reserved. May not be scanned, copied or duplicated, or posted to a publicly accessible website, in whole or in part.

3.  The test is inconclusive if $\lim\limits_{n\to\infty} \sqrt[n]{|a_n|}$ _____.

## III.  Strategies for Testing Series  (Pages 645–646)

List four guidelines for testing a series for convergence or
divergence.

What you should learn<br/>
How to review the tests
for convergence and
divergence of an infinite
series

1.

2.

3.

4.

Complete the following selected tests for series.

| Test | Series | Converges | Diverges |
|------|--------|-----------|----------|
| $n$th-Term | $\sum\limits_{n=1}^{\infty} a_n$ | | |
| | $\sum\limits_{n=0}^{\infty} ar^n$ | | |
| | $\sum\limits_{n=1}^{\infty} \dfrac{1}{n^p}$ | | |
| | $\sum\limits_{n=1}^{\infty} (b_n - b_{n+1})$ | | |
| Ratio | $\sum\limits_{n=1}^{\infty} a_n$ | | |

---

**Homework Assignment**

Page(s)

Exercises

---

© 2011 Cengage Learning. All Rights Reserved. May not be scanned, copied or duplicated, or posted to a publicly accessible website, in whole or in part.

# Section 9.7  Taylor Polynomials and Approximations

**Objective:**    In this lesson you learned how to find Taylor or
Maclaurin polynomial approximations of elementary
functions.

Course Number

Instructor

Date

## I. Polynomial Approximations of Elementary Functions
(Pages 650–651)

To find a polynomial function $P$ that approximates another

function $f$, _____

_____

_____. The approximating polynomial is

said to be _____.

*What you should learn*
How to find polynomial
approximations of
elementary functions and
compare them with the
elementary functions

## II. Taylor and Maclaurin Polynomials  (Pages 652–655)

If $f$ has $n$ derivatives at $c$, then the polynomial

$$P_n(x) = f(c) + f'(c)(x-c) + \frac{f''(c)}{2!}(x-c)^2 + \cdots + \frac{f^{(n)}(c)}{n!}(x-c)^n$$

is called the _____.

If $c = 0$, then $P_n(x) = f(0) + f'(0)x + \frac{f''(0)}{2!}x^2 + \cdots + \frac{f^{(n)}(0)}{n!}x^n$

is also called the _____.

The accuracy of a Taylor or Maclaurin polynomial

approximation is usually better at $x$-values _____,

_____. The approximation is

usually better for higher-degree Taylor or Maclaurin

polynomials than _____.

*What you should learn*
How to find Taylor and
Maclaurin polynomial
approximations of
elementary functions

## III.  Remainder of a Taylor Polynomial  (Pages 656–657)

If a function $f$ is differentiable through order $n + 1$ in an interval $I$

containing $c$, then for each $x$ in $I$, Taylor's Theorem states that

there exists $z$ between $x$ and $c$ such that $f(x) =$ _____

_____,

*What you should learn*
How to use the remainder
of a Taylor polynomial

© 2011 Cengage Learning. All Rights Reserved. May not be scanned, copied or duplicated, or posted to a publicly accessible website, in whole or in part.

where $R_n(x)$ is given by $R_n(x) = \dfrac{f^{(n+1)}(z)}{(n+1)!}(x-c)^{n+1}$. The value

$R_n(x)$ is called the _____.

The practical application of this theorem lies not in calculating

$R_n(x)$, but in _____

_____ .

---

**Homework Assignment**

Page(s)

Exercises

---

© 2011 Cengage Learning. All Rights Reserved. May not be scanned, copied or duplicated, or posted to a publicly accessible website, in whole or in part.

## Section 9.8  Power Series

**Objective:**    In this lesson you learned how to find the radius and interval of convergence of power series and how to differentiate and integrate power series.

Course Number

Instructor

Date

### I. Power Series  (Pages 661–662)

If $x$ is a variable, then an infinite series of the form

$$\sum_{n=0}^{\infty} a_n x^n = a_0 + a_1 x + a_2 x^2 + a_3 x^3 + \cdots + a_n x^n + \cdots \text{ is called a}$$

_____. More generally, an

infinite series of the form

$$\sum_{n=0}^{\infty} a_n (x-c)^n = a_0 + a_1(x-c) + a_2(x-c)^2 + \cdots + a_n(x-c)^n + \cdots$$

is called a _____,

where $c$ is a constant.

> **What you should learn**
> How to understand the definition of a power series

### II.  Radius and Interval of Convergence  (Pages 662–663)

For a power series centered at $c$, precisely one of the following is true.

1.  The series converges only at _____.

2.  There exists a real number $R > 0$ such that the series converges absolutely for _____, and diverges for _____.

3.  The series converges absolutely for _____.

The number $R$ is the _____ of the power series. If the series converges only at $c$, the radius of convergence is _____, and if the series converges for all $x$, the radius of convergence is _____.

The set of all values of $x$ for which the power series converges is the _____ of the power series.

> **What you should learn**
> How to find the radius and interval of convergence of a power series

© 2011 Cengage Learning. All Rights Reserved. May not be scanned, copied or duplicated, or posted to a publicly accessible website, in whole or in part.

### III. Endpoint Convergence  (Pages 664–665)

**What you should learn**
How to determine the
endpoint convergence of
a power series

For a power series whose radius of convergence is a finite

number $R$, each endpoint of the interval of convergence must be

_____ .


### IV.  Differentiation and Integration of Power Series
(Pages 666–667)

**What you should learn**
How to differentiate and
integrate a power series

If the function given by

$$f(x) = \sum_{n=0}^{\infty} a_n (x-c)^n$$

$$= a_0 + a_1(x-c) + a_2(x-c)^2 + a_3(x-c)^3 + \cdots$$

has a radius of convergence of $R > 0$, then, on the interval

$(c-R, c+R), f$ is _____

_____ . Moreover, the derivative and antiderivative

of $f$ are as follows.

1. $f'(x) = \displaystyle\sum_{n=1}^{\infty} n a_n (x-c)^{n-1}$

   $=$ _____

2. $\displaystyle\int f(x)\, dx = C + \sum_{n=0}^{\infty} a_n \frac{(x-c)^{n+1}}{n+1}$

   $=$ _____


The radius of convergence of the series obtained by

differentiating or integrating a power series is _____

_____ . The interval

of convergence, however, may differ as a result of _____

_____ .


**Homework Assignment**

Page(s)

Exercises

© 2011 Cengage Learning. All Rights Reserved. May not be scanned, copied or duplicated, or posted to a publicly accessible website, in whole or in part.

## Section 9.9   Representation of Functions by Power Series

**Objective:**      In this lesson you learned how to represent functions by
power series.

Course Number

Instructor

Date

### I. Geometric Power Series  (Pages 671–672)

Describe two ways for finding a geometric power series.

*What you should learn*
How to find a geometric
power series that
represents a function

### II. Operations with Power Series  (Pages 673–675)

Let $f(x) = \sum a_n x^n$ and $g(x) = \sum b_n x^n$ .

*What you should learn*
How to construct a power
series using series
operations

1. $f(kx) = \displaystyle\sum_{n=0}^{\infty}$ _____

2. $f(x^N) = \displaystyle\sum_{n=0}^{\infty}$ _____

3. $f(x) \pm g(x) = \displaystyle\sum_{n=0}^{\infty}$ _____

The operations described above can change _____

_____ .

© 2011 Cengage Learning. All Rights Reserved. May not be scanned, copied or duplicated, or posted to a publicly accessible website, in whole or in part.

**Additional notes**

**Homework Assignment**

Page(s)

Exercises

© 2011 Cengage Learning. All Rights Reserved. May not be scanned, copied or duplicated, or posted to a publicly accessible website, in whole or in part.

## Section 9.10  Taylor and Maclaurin Series

**Objective:**   In this lesson you learned how to find a Taylor or
Maclaurin series for a function.

**Course Number**

**Instructor**

**Date**

**I. Taylor Series and Maclaurin Series**  (Pages 678–682)

**The Form of a Convergent Power Series**

*What you should learn*
How to find a Taylor or
Maclaurin series for a
function

If $f$ is represented by a power series $f(x) = \sum a_n(x-c)^n$ for all

$x$ in an open interval $I$ containing $c$, then $a_n =$ _____,

and $f(x) =$ _____

_____.

The series is called the **Taylor series** for $f(x)$ at $c$ because _____

_____

_____.

If a function $f$ has derivatives of all orders at $x = c$, then the

series $\sum_{n=0}^{\infty} \dfrac{f^{(n)}(c)}{n!}(x-c)^n =$ _____

_____ is called

the **Taylor series for $f(x)$ at $c$.** Moreover, if $c = 0$, then the series

is called the _____.

If $\lim\limits_{n \to \infty} R_n = 0$ for all $x$ in the interval $I$, then the Taylor series for $f$

_____, where

$f(x) = \sum_{n=0}^{\infty} \dfrac{f^{(n)}(c)}{n!}(x-c)^n$ .

Complete the list of guidelines for finding a Taylor series.

1.

© 2011 Cengage Learning. All Rights Reserved. May not be scanned, copied or duplicated, or posted to a publicly accessible website, in whole or in part.

2.


3.


## II. Binomial Series  (Page 683)

The **binomial series** for a function of the form $f(x) = (1+x)^k$ is

_____

*What you should learn*
How to find a binomial
series

## III. Deriving Taylor Series from a Basic List
   (Pages 684–686)

Because direct computation of Taylor or Maclaurin coefficients
can be tedious, the most practical way to find a Taylor or
Maclaurin series is to develop power series for a basic list of
elementary functions. From this list, you can determine power
series for other functions by the operations of _____

_____

_____with known
power series.

*What you should learn*
How to use a basic list of
Taylor series to find other
Taylor series

List power series for the following elementary functions and
give the interval of convergence for each.


$\dfrac{1}{x} =$


$\dfrac{1}{1+x} =$


$\ln x =$


$e^x =$

© 2011 Cengage Learning. All Rights Reserved. May not be scanned, copied or duplicated, or posted to a publicly accessible website, in whole or in part.

$\sin x =$

$\cos x =$

$\arctan x =$

$\arcsin x =$

$(1 + x)^k =$

**Additional notes**

© 2011 Cengage Learning. All Rights Reserved. May not be scanned, copied or duplicated, or posted to a publicly accessible website, in whole or in part.

**Additional notes**

**Homework Assignment**

Page(s)

Exercises

Larson/Edwards  **Calculus:  Early Transcendental Functions 5e**  Notetaking Guide

© 2011 Cengage Learning. All Rights Reserved. May not be scanned, copied or duplicated, or posted to a publicly accessible website, in whole or in part.

# Chapter 10    Conics, Parametric Equations, And Polar Coordinates

| |
|---|
| Course Number |
| Instructor |
| Date |

## Section 10.1    Conics and Calculus

**Objective:** In this lesson you learned how to analyze and write an equation of a parabola, an ellipse, and a hyperbola.

---

**Important Vocabulary**        Define each term or concept.

**Directrix of a parabola**


**Focus of a parabola**

**Tangent of parabola**

**Foci of an ellipse**

**Vertices of an ellipse**

**Major axis of an ellipse**

**Center of an ellipse**

**Minor axis of an ellipse**

**Branches of a hyperbola**

**Transverse axis of a hyperbola**

**Conjugate axis of a hyperbola**

---

**I.  Conic Sections**  (Page 696)

A **conic section,** or **conic,** is _____

_____ .

| |
|---|
| **What you should learn** |
| Understand the definition of a conic section |

Name the four basic conic sections: _____

_____ .

In the formation of the four basic conics, the intersecting plane

does not pass through the vertex of the cone. When the plane

does pass through the vertex, the resulting figure is a(n)

© 2011 Cengage Learning. All Rights Reserved. May not be scanned, copied or duplicated, or posted to a publicly accessible website, in whole or in part.

_____ , such as

_____

_____ .

In this section, each conic is defined as a _____ of

points satisfying a certain geometric property. For example, a

circle is the collection of all points $(x, y)$ that are

_____ from a fixed point $(h, k)$. This

locus definition easily produces the standard equation of a circle

_____ .

## II. Parabolas  (Pages 697–698)

A **parabola** is _____

_____

_____ .

| What you should learn |
| How to analyze and write equations of parabolas using properties of parabolas |

The midpoint between the focus and the directrix is the

_____ of a parabola. The line passing through the

focus and the vertex is the _____ of the parabola.

The **standard form** of the equation of a parabola with a vertical

axis having a vertex at $(h, k)$ and directrix $y = k - p$ is

_____

The standard form of the equation of a parabola with a horizontal

axis having a vertex at $(h, k)$ and directrix $x = h - p$ is

_____

The focus lies on the axis $p$ units (directed distance) from the

vertex. The coordinates of the focus are _____ for

a vertical axis or _____ for a horizontal axis.

**Example 1:**   Find the standard form of the equation of the
                 parabola with vertex at the origin and focus $(1, 0)$.

© 2011 Cengage Learning. All Rights Reserved. May not be scanned, copied or duplicated, or posted to a publicly accessible website, in whole or in part.

A **focal chord** is _____

_____.

The specific focal chord perpendicular to the axis of a parabola

is called the _____.

The reflective property of a parabola states that the tangent line
to a parabola at a point $P$ makes equal angles with the following
two lines:

1)

2)

### III.  Ellipses  (Pages 699–702)

An **ellipse** is _____

_____

_____.

The standard form of the equation of an ellipse with center $(h, k)$

and a horizontal major axis of length $2a$ and a minor axis of

length $2b$, where $a > b$, is: _____

The standard form of the equation of an ellipse with center $(h, k)$

and a vertical major axis of length $2a$ and a minor axis of length

$2b$, where $a > b$, is: _____

In both cases, the foci lie on the major axis, $c$ units from the

center, with $c^2 =$ _____.

> **What you should learn**
> How to analyze and write
> equations of ellipses
> using properties of
> ellipses

**Example 2:**  Sketch the ellipse given by $4x^2 + 25y^2 = 100$ .

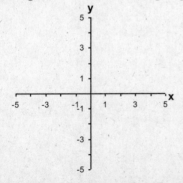

© 2011 Cengage Learning. All Rights Reserved. May not be scanned, copied or duplicated, or posted to a publicly accessible website, in whole or in part.

Let $P$ be a point on an ellipse. The Reflective Property of an Ellipse states that _____

_____

_____.


_____ measures the ovalness of an ellipse. It is given by the ratio $e = $ _____. For an elongated ellipse, the value of $e$ is close to _____. For every ellipse, the value of $e$ lies between _____ and _____.


**IV.  Hyperbolas**  (Pages 703–705)

A **hyperbola** is _____

_____

_____.

| What you should learn |
| --- |
| How to analyze and write equations of hyperbolas using properties of hyperbolas |

The line through a hyperbola's two foci intersects the hyperbola at two points called _____.

The midpoint of a hyperbola's transverse axis is the _____ of the hyperbola.

The standard form of the equation of a hyperbola centered at $(h, k)$ and having a horizontal transverse axis is

_____

The standard form of the equation of a hyperbola centered at $(h, k)$ and having a vertical transverse axis is

_____

The vertices are $a$ units from the center and the foci are $c$ units from the center. Moreover, $a$, $b$, and $c$ are related by the equation

_____.

The **asymptotes** of a hyperbola with a horizontal transverse axis are _____.

The **asymptotes** of a hyperbola with a vertical transverse axis are _____.

© 2011 Cengage Learning. All Rights Reserved. May not be scanned, copied or duplicated, or posted to a publicly accessible website, in whole or in part.

**Example 3:**    Sketch the graph of the hyperbola given by

$$y^2 - 9x^2 = 9.$$

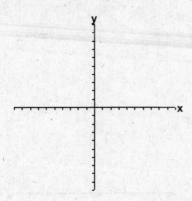

The **eccentricity** of a hyperbola is $e =$ _____, where

the values of $e$ are _____.

**Additional notes**

© 2011 Cengage Learning. All Rights Reserved. May not be scanned, copied or duplicated, or posted to a publicly accessible website, in whole or in part.

**Additional notes**

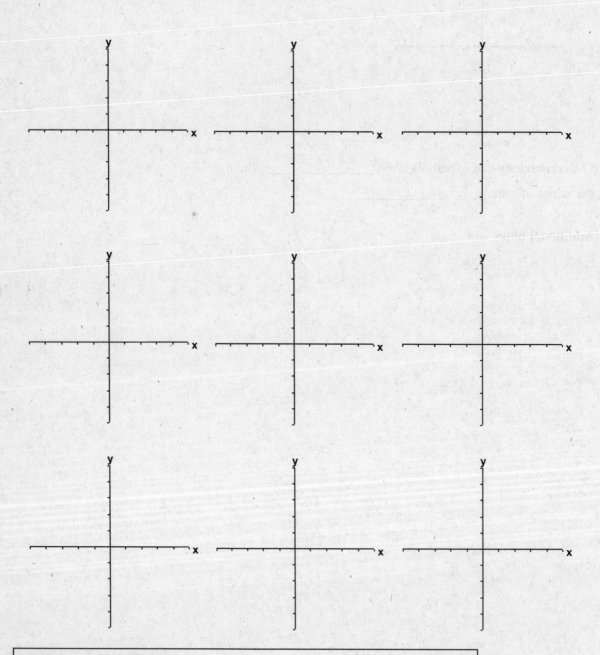

**Homework Assignment**

Page(s)

Exercises

© 2011 Cengage Learning. All Rights Reserved. May not be scanned, copied or duplicated, or posted to a publicly accessible website, in whole or in part.

Section 10.2  Plane Curves and Parametric Equations                                    **189**

## Section 10.2  Plane Curves and Parametric Equations

**Course Number**

**Instructor**

**Date**

**Objective:**  In this lesson you learned how to sketch a curve represented by parametric equations.

### I. Plane Curves and Parametric Equations  (Pages 711–712)

If $f$ and $g$ are continuous functions of $t$ on an interval $I$, then the equations $x = f(t)$ and $y = g(t)$ are called _____ _____ and $t$ is called the _____.

The set of points $(x, y)$ obtained as $t$ varies over the interval $I$ is called the _____.

Taken together, the parametric equations and the graph are called a _____, denoted by $C$.

When sketching (by hand) a curve represented by a set of parametric equations, you can  plot points in the _____.

Each set of coordinates $(x, y)$ is determined from a value chosen for the _____. By plotting the resulting points in the order of increasing values of $t$, the curve is traced out in a specific direction, called the _____ of the curve.

*What you should learn*
How to sketch the graph of a curve given by a set of parametric equations

**Example 1:**  Sketch the curve described by the parametric equations $x = t - 3$ and $y = t^2 + 1$, $-1 \le t \le 3$.

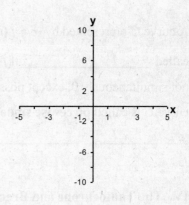

### II. Eliminating the Parameter  (Pages 713–714)

**Eliminating the parameter** is the process of _____ _____ _____.

Describe the process used to eliminate the parameter from a set of parametric equations.

*What you should learn*
How to eliminate the parameter in a set of parametric equations

Larson/Edwards  **Calculus:  Early Transcendental Functions 5e**  Notetaking Guide

© 2011 Cengage Learning. All Rights Reserved. May not be scanned, copied or duplicated, or posted to a publicly accessible website, in whole or in part.

When converting equations from parametric to rectangular form, the range of $x$ and $y$ implied by the parametric equations may be _____ by the change to rectangular form. In such instances, the domain of the rectangular equation must be

_____

_____.

To eliminate the parameter in equations involving trigonometric functions, try using the identity _____.

## III.  Finding Parametric Equations  (Pages 715–716)

Describe how to find a set of parametric equations for a given graph.

*What you should learn*
How to find a set of parametric equations to represent a curve

A curve $C$ represented by $x = f(t)$ and $y = g(t)$ on an interval $I$ is called _____ if $f'$ and $g'$ are continuous on $I$ and not simultaneously 0, except possibly at the endpoints of $I$. The curve $C$ is called **piecewise smooth** if _____

_____.

## IV.  The Tautochrone and Brachistochrone Problems
     (Page 717)

Describe the tautochrone problem and the brachistochrone problem in your own words.

*What you should learn*
Understand two classic calculus problems, the tautochrone and brachistochrone problems

**Homework Assignment**

Page(s)

Exercises

© 2011 Cengage Learning. All Rights Reserved. May not be scanned, copied or duplicated, or posted to a publicly accessible website, in whole or in part.

## Section 10.3    Parametric Equations and Calculus

Course Number

Instructor

Date

**Objective:**    In this lesson you learned how to use a set of parametric equations to find the slope of a tangent line to a curve and the arc length of a curve.

### I. Slope and Tangent Lines  (Pages 721–723)

If a smooth curve $C$ is given by the equations $x = f(t)$ and

$y = g(t)$, then the slope of $C$ at $(x, y)$ is $\dfrac{dy}{dx} =$ _____ ,

$\dfrac{dx}{dt} \neq$ _____ .

*What you should learn*
How to find the slope of a tangent line to a curve by a set of parametric equations

**Example 1:**    For the curve given by the parametric equations
$x = t - 3$ and $y = t^2 + 1$, $-1 \leq t \leq 3$, find the slope at the point $(-3, 1)$.

### II. Arc Length  (Pages 723–725)

If a smooth curve $C$ is given by $x = f(t)$ and $y = g(t)$ such that $C$ does not intersect itself on the interval $a \leq t \leq b$ (except possibly at the endpoints), then the arc length of $C$ over the interval is given by

$$s = \int_a^b \sqrt{\rule{2cm}{0pt}}\; dt = \int_a^b \sqrt{\rule{2cm}{0pt}}\; dt$$

*What you should learn*
How to find the arc length of a curve given by a set of parametric equations

In the preceding section you saw that if a circle rolls along a line, a point on its circumference will trace a path called a

_____ . If the circle rolls around the

circumference of another circle, the path of the point is an

_____ .

© 2011 Cengage Learning. All Rights Reserved. May not be scanned, copied or duplicated, or posted to a publicly accessible website, in whole or in part.

### III. Area of Surface of Revolution  (Page 726)

**What you should learn**
How to find the area of a
surface of revolution
(parametric form)

If a smooth curve $C$ given by $x = f(t)$ and $y = g(t)$ does not
cross itself on the interval $a \leq t \leq b$, then the area $S$ of the
surface of revolution formed by revolving $C$ about the coordinate
axes is given by

1.  $S = \displaystyle\int \sqrt{\phantom{xxxxxxxxxx}}$ 　　　　Revolution about the _____: $g(t) \geq 0$

2.  $S = \displaystyle\int \sqrt{\phantom{xxxxxxxxxx}}$ 　　　　Revolution about the _____: $f(t) \geq 0$

**Homework Assignment**

Page(s)

Exercises

© 2011 Cengage Learning. All Rights Reserved. May not be scanned, copied or duplicated, or posted to a publicly accessible website, in whole or in part.

## Section 10.4    Polar Coordinates and Polar Graphs

**Objective:**    In this lesson you learned how to sketch the graph of an equation in polar form, find the slope of a tangent line to a polar graph, and identify special polar graphs.

---

Course Number

Instructor

Date

---

### I.  Polar Coordinates  (Page 731)

To form the **polar coordinate system** in the plane, fix a point $O$, called the _____ or _____, and construct from $O$ an initial ray called the _____. Then each point $P$ in the plane can be assigned _____ $(r, \theta)$ as follows:

1) $r =$ _____

2) $\theta =$ _____

_____

In the polar coordinate system, points do not have a unique representation. In general, the point $(r, \theta)$ can be represented as

_____ or _____, where

$n$ is any integer. Moreover, the pole is represented by $(0, \theta)$, where $\theta$ is _____.

---

> **What you should learn**
> How to understand the polar coordinate system

---

**Example 1:**    Plot the point $(r, \theta) = (-2, 11\pi/4)$ on the polar coordinate system.

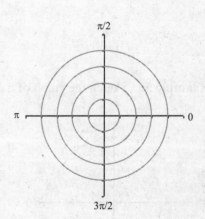

**Example 2:**    Find another polar representation of the point $(4, \pi/6)$.

---

© 2011 Cengage Learning. All Rights Reserved. May not be scanned, copied or duplicated, or posted to a publicly accessible website, in whole or in part.

## II. Coordinate Conversion (Page 732)

The polar coordinates $(r, \theta)$ of a point are related to the
rectangular coordinates $(x, y)$ of the point as follows . . .

*What you should learn*
How to rewrite
rectangular coordinates
and equations in polar
form and vice versa

**Example 3:**  Convert the polar coordinates $(3, 3\pi/2)$ to
rectangular coordinates.

## III. Polar Graphs (Pages 733–734)

One way to sketch the graph of a polar equation is to

_____

_____ .

*What you should learn*
How to sketch the graph
of an equation given in
polar form

To convert a rectangular equation to polar form, _____

_____ .

**Example 4:**  Find the rectangular equation corresponding to the

polar equation $r = \dfrac{-5}{\sin\theta}$.

**Example 5:**  Sketch the graph of the polar equation $r = 3\cos\theta$.

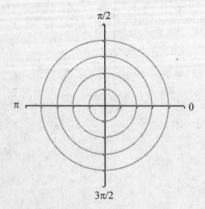

© 2011 Cengage Learning. All Rights Reserved. May not be scanned, copied or duplicated, or posted to a publicly accessible website, in whole or in part.

**IV. Slope and Tangent Lines** (Pages 735–736)

*What you should learn*
How to find the slope of
a tangent line to a polar
graph

If $f$ is a differentiable function of $\theta$, then the slope of the tangent line to the graph of $r = f(\theta)$ at the point $(r, \theta)$ is

$$\frac{dy}{dx} = \frac{dy/d\theta}{dx/d\theta} = \underline{\hspace{4cm}},$$

provided that $\dfrac{dx}{d\theta} \neq 0$ at $(r, \theta)$.

Solutions to $\dfrac{dy}{d\theta} = 0$ yield $\underline{\hspace{5cm}}$,

provided that $\dfrac{dx}{d\theta} \neq 0$. Solutions to $\dfrac{dx}{d\theta} = 0$ yield $\underline{\hspace{2cm}}$

$\underline{\hspace{3cm}}$, provided that $\dfrac{dy}{d\theta} \neq 0$.

If $f(\alpha) = 0$ and $f'(\alpha) \neq 0$, then the line $\theta = \alpha$ is $\underline{\hspace{2cm}}$

$\underline{\hspace{6cm}}$. This theorem

is useful because it states that $\underline{\hspace{4cm}}$

$\underline{\hspace{6cm}}$.

**V. Special Polar Graphs** (Page 737)

*What you should learn*
How to identify several
types of special polar
graphs

List the general equations that yield each of the following types of special polar graphs:

Limaçons:

Rose curves:

Circles:

Lemniscates:

© 2011 Cengage Learning. All Rights Reserved. May not be scanned, copied or duplicated, or posted to a publicly accessible website, in whole or in part.

**Additional notes**

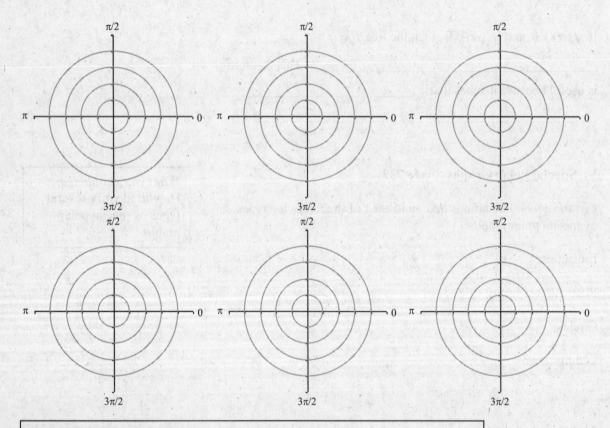

**Homework Assignment**

Page(s)

Exercises

© 2011 Cengage Learning. All Rights Reserved. May not be scanned, copied or duplicated, or posted to a publicly accessible website, in whole or in part.

# Section 10.5    Area and Arc Length in Polar Coordinates

**Objective:**    In this lesson you learned how to find the area of a
region bounded by a polar graph and the arc length of a
polar graph.

Course Number

Instructor

Date

## I.  Area of a Polar Region  (Pages 741–742)

If $f$ is continuous and nonnegative on the interval $[\alpha, \beta]$,

$0 < \beta - \alpha \le 2\pi$, then the area of the region bounded by the graph

of $r = f(\theta)$ between the radial lines $\theta = \alpha$ and $\theta = \beta$ is given

by _____ .

*What you should learn*
How to find the area of a
region bounded by a
polar graph

## II.  Points of Intersection of Polar Graphs  (Pages 743–744)

Explain why care must be taken in determining the points of
intersection of two polar graphs.

*What you should learn*
How to find the points of
intersection of two polar
graphs

## III.  Arc Length in Polar Form  (Page 745)

Let $f$ be a function whose derivative is continuous on an interval

$\alpha \le \theta \le \beta$. The length of the graph of $r = f(\theta)$ from $\theta = \alpha$ to

$\theta = \beta$ is _____ .

*What you should learn*
How to find the arc
length of a polar graph

© 2011 Cengage Learning. All Rights Reserved. May not be scanned, copied or duplicated, or posted to a publicly accessible website, in whole or in part.

**IV. Area of a Surface of Revolution** (Page 746)

Let $f$ be a function whose derivative is continuous on an interval $\alpha \le \theta \le \beta$. The area of the surface formed by revolving the graph of $r = f(\theta)$ from $\theta = \alpha$ to $\theta = \beta$ about the indicated line is as follows.

*What you should learn*
How to find the area of a surface of revolution (polar form)

1.  About the polar axis:

    _____

2.  About the line $\theta = \dfrac{\pi}{2}$ :

    _____

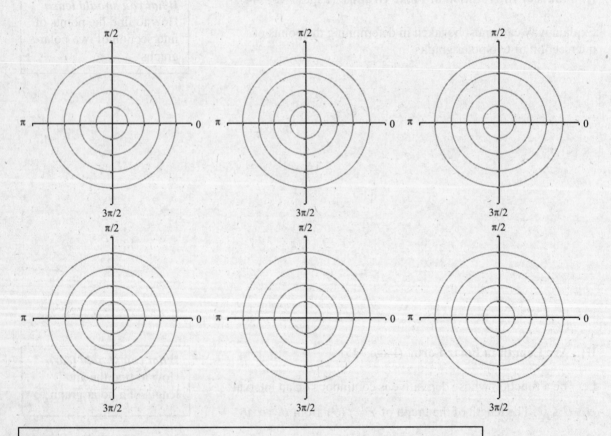

**Homework Assignment**

Page(s)

Exercises

© 2011 Cengage Learning. All Rights Reserved. May not be scanned, copied or duplicated, or posted to a publicly accessible website, in whole or in part.

## Section 10.6   Polar Equations of Conics and Kepler's Laws

Course Number

Instructor

Date

**Objective:**   In this lesson you learned how to analyze and write a
polar equation of a conic.

**I. Polar Equations of Conics**  (Pages 750–752)

*What you should learn*
How to analyze and write
polar equations of conics

Let $F$ be a fixed point (*focus*) and $D$ be a fixed line (*directrix*) in
the plane. Let $P$ be another point in the plane and let $e$
(*eccentricity*) be the ratio of the distance between $P$ and $F$ to the
distance between $P$ and $D$. The collection of all points $P$ with a
given eccentricity is a _____.

The conic is an ellipse if _____. The conic is a
parabola if _____. Finally, the conic is a hyperbola
if _____.

For each type of conic, the pole corresponds to the _____.

The graph of the polar equation _____
is a conic with a vertical directrix to the right of the pole, where
$e > 0$ is the eccentricity and $|\,d\,|$ is the distance between the focus
(pole) and the directrix.

The graph of the polar equation _____
is a conic with a vertical directrix to the left of the pole, where
$e > 0$ is the eccentricity and $|\,d\,|$ is the distance between the focus
(pole) and the directrix.

The graph of the polar equation _____
is a conic with a horizontal directrix above the pole, where $e > 0$
is the eccentricity and $|\,d\,|$ is the distance between the focus
(pole) and the directrix.

The graph of the polar equation _____
is a conic with a horizontal directrix below the pole, where $e > 0$
is the eccentricity and $|\,d\,|$ is the distance between the focus
(pole) and the directrix.

© 2011 Cengage Learning. All Rights Reserved. May not be scanned, copied or duplicated, or posted to a publicly accessible website, in whole or in part.

**Example 1:**  Identify the type of conic from the polar equation
$$r = \frac{36}{10 + 12\sin\theta}, \text{ and describe its orientation.}$$

## II. Kepler's Laws  (Pages 753–754)

List Kepler's Laws, which can be used to describe the orbits of the planets about the sun.

1.

2.

3.

> **What you should learn**
> How to understand and use Kepler's Laws of planetary motion

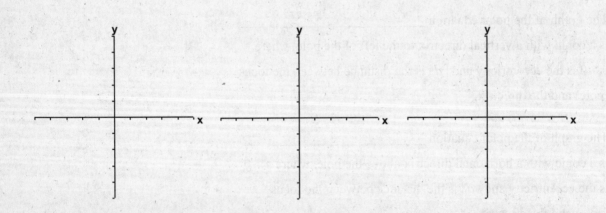

---

**Homework Assignment**

Page(s)

Exercises

---

© 2011 Cengage Learning. All Rights Reserved. May not be scanned, copied or duplicated, or posted to a publicly accessible website, in whole or in part.

# Chapter 11   Vectors and the Geometry of Space

<table>
<tr><td>Course Number</td></tr>
<tr><td>Instructor</td></tr>
<tr><td>Date</td></tr>
</table>

## Section 11.1      Vectors in the Plane

**Objective:** In this lesson you learned how to represent vectors, perform basic vector operations, and represent vectors graphically.

---

**Important Vocabulary**      Define each term or concept.

**Vector v in the plane**

**Standard position**

**Zero vector**

**Unit vector**

**Standard unit vectors**

---

## I. Component Form of a Vector  (Pages 764–765)

> **What you should learn**
> How to write the component form of a vector

A **directed line segment** has an _____ and a
_____.

The **magnitude** of the directed line segment $\overrightarrow{PQ}$, denoted by
_____, is its _____. The length of a
directed line segment can be found by _____
_____.

If **v** is a vector in the plane whose initial point is at the origin and
whose terminal point is $(v_1, v_2)$, then the _____
_____ is given by $\mathbf{v} = \langle v_1, v_2 \rangle$, where the
coordinates $v_1$ and $v_2$ are called the _____.

If $P\,(p_1, p_2)$ and $Q\,(q_1, q_2)$ are the initial and terminal points of a
directed line segment, the component form of the vector **v**
represented by $\overrightarrow{PQ}$ is _____ = _____.

The **length** (or magnitude) **of v** is:
$$\|\mathbf{v}\| = \sqrt{\rule{4cm}{0pt}} = \sqrt{\rule{2cm}{0pt}}$$

---

© 2011 Cengage Learning. All Rights Reserved. May not be scanned, copied or duplicated, or posted to a publicly accessible website, in whole or in part.

If $\mathbf{v} = \langle v_1, v_2 \rangle$, $\mathbf{v}$ can be represented by the _____

_____ from $P(0, 0)$ to

$Q(v_1, v_2)$.

The length of $\mathbf{v}$ is also called the _____.

**Example 1:**  Find the component form and length of the vector
$\mathbf{v}$ that has (1, 7) as its initial point and (4, 3) as its
terminal point.

## II.  Vector Operations  (Pages 766–769)

Let $\mathbf{u} = \langle u_1, u_2 \rangle$ and $\mathbf{v} = \langle v_1, v_2 \rangle$ be vectors and let $c$ be a scalar.

Then the **vector sum** of $\mathbf{u}$ and $\mathbf{v}$ is the vector:

   $\mathbf{u} + \mathbf{v} =$ _____

and the **scalar multiple** of $c$ and $\mathbf{u}$ is the vector:

   $c\mathbf{u} =$ _____.

Furthermore, the **negative** of $\mathbf{v}$ is the vector

   $-\mathbf{v} =$ _____

and the **difference** of $\mathbf{u}$ and $\mathbf{v}$ is

   $\mathbf{u} - \mathbf{v} =$ _____

Geometrically, the scalar multiple of a vector $\mathbf{v}$ and a scalar $c$ is

_____.

If $c$ is positive, $c\mathbf{v}$ has the _____ direction as $\mathbf{v}$, and if $c$
is negative, $c\mathbf{v}$ has the _____ direction.

To add two vectors geometrically, _____

_____

_____.

The vector $\mathbf{u} + \mathbf{v}$, called the _____, is

_____

_____.

> **What you should learn**
> How to perform vector
> operations and interpret
> the results geometrically

© 2011 Cengage Learning. All Rights Reserved. May not be scanned, copied or duplicated, or posted to a publicly accessible website, in whole or in part.

**Example 2:**   Let $\mathbf{u} = \langle 1, 6 \rangle$ and $\mathbf{v} = \langle -4, 2 \rangle$. Find:
   (a)  $3\mathbf{u}$          (b)  $\mathbf{u} + \mathbf{v}$

Let $\mathbf{u}$, $\mathbf{v}$, and $\mathbf{w}$ be vectors in the plane, and let $c$ and $d$ be scalars.
Complete the following properties of vector addition and scalar
multiplication:

1.  $\mathbf{u} + \mathbf{v} = $ _____

2.  $(\mathbf{u} + \mathbf{v}) + \mathbf{w} = $ _____

3.  $\mathbf{u} + \mathbf{0} = $ _____

4.  $\mathbf{u} + (-\mathbf{u}) = $ _____

5.  $c(d\mathbf{u}) = $ _____

6.  $(c + d)\mathbf{u} = $ _____

7.  $c(\mathbf{u} + \mathbf{v}) = $ _____

8.  $1(\mathbf{u}) = $ _____ ;   $0(\mathbf{u}) = $ _____

Any set of vectors, with an accompanying set of scalars, that

satisfies these eight properties is a _____.

Let $\mathbf{v}$ be a vector and let $c$ be a scalar. Then

$\|c\mathbf{v}\| = $ _____

To find a unit vector $\mathbf{u}$ that has the same direction as a given

nonzero vector $\mathbf{v}$, _____

_____.

In this case, the vector $\mathbf{u}$ is called a _____

_____. The process of multiplying $\mathbf{v}$ by $1/\|\mathbf{v}\|$

to get a unit vector is called _____.

**Example 3:**   Find a unit vector in the direction of $\mathbf{v} = \langle -8, 6 \rangle$.

© 2011 Cengage Learning. All Rights Reserved. May not be scanned, copied or duplicated, or posted to a publicly accessible website, in whole or in part.

### III. Standard Unit Vectors (Pages 769–770)

*What you should learn*
How to write a vector as
a linear combination of
standard unit vectors

Let $\mathbf{v} = \langle v_1, v_2 \rangle$. Then the standard unit vectors can be used to

represent $\mathbf{v}$ as $\mathbf{v} = $ _____ , where the scalar $v_1$ is

called the _____ and the scalar

$v_2$ is called the _____. The vector

sum $v_1 \mathbf{i} + v_2 \mathbf{j}$ is called a _____ of the

vectors $\mathbf{i}$ and $\mathbf{j}$.

**Example 4:** Let $\mathbf{v} = \langle -5, 3 \rangle$. Write $\mathbf{v}$ as a linear combination of
the standard unit vectors $\mathbf{i}$ and $\mathbf{j}$.

**Example 5:** Let $\mathbf{v} = 3\mathbf{i} - 4\mathbf{j}$ and $\mathbf{w} = 2\mathbf{i} + 9\mathbf{j}$. Find $\mathbf{v} + \mathbf{w}$.

If $\mathbf{u}$ is a unit vector and $\theta$ is the angle (measured

counterclockwise) from the positive $x$-axis to $\mathbf{u}$, the terminal

point of $\mathbf{u}$ lies on the unit circle and $\mathbf{u} = $ _____ =

_____.

Now, if $\mathbf{v}$ is any nonzero vector that makes an angle $\theta$ with the

positive $x$-axis, it has the same direction as $\mathbf{u}$ and

$\mathbf{v} = $ _____ = _____.

### IV. Applications of Vectors (Pages 770–771)

*What you should learn*
How to use vectors to
solve problems involving
force or velocity

Describe several real-life applications of vectors.

**Homework Assignment**

Page(s)

Exercises

© 2011 Cengage Learning. All Rights Reserved. May not be scanned, copied or duplicated, or posted to a publicly accessible website, in whole or in part.

## Section 11.2   Space Coordinates and Vectors in Space

Course Number

Instructor

Date

**Objective:**   In this lesson you learned how to plot points in a three-dimensional coordinate system and analyze vectors in space.

---

**Important Vocabulary**          Define each term or concept.

**Sphere**

**Standard unit vector notation in space**

**Parallel vectors in space**

---

### I.  Coordinates in Space  (Pages 775–776)

A **three-dimensional coordinate system** is constructed by
_____
_____.

*What you should learn*
How to understand the three-dimensional rectangular coordinate system

Taken as pairs, the axes determine three coordinate planes:  the
_____, the _____, and the _____.

These three coordinate planes separate the three-space into eight
_____. The first of these is the one for which ___
_____.

In the three-dimensional system, a point $P$ in space is determined by an ordered triple $(x, y, z)$, where $x$, $y$, and $z$ are as follows . . .

$x = $ _____,

$y = $ _____,

and $z = $ _____.

A three-dimensional coordinate system can have either a _____
_____ or a _____ orientation. To
determine the orientation of a system, _____
_____

---

Larson/Edwards  **Calculus: Early Transcendental Functions 5e**  Notetaking Guide

© 2011 Cengage Learning. All Rights Reserved. May not be scanned, copied or duplicated, or posted to a publicly accessible website, in whole or in part.

_____

_____

_____.

The distance between the points $(x_1, y_1, z_1)$ and $(x_2, y_2, z_2)$ given by the Distance Formula in space is

$$d = \sqrt{\phantom{xxxxxxxxxxxxxxxxxxxxxxxxxxxxxxxxxx}}$$

The midpoint of the line segment joining the points $(x_1, y_1, z_1)$ and $(x_2, y_2, z_2)$ given by the **Midpoint Formula in Space** is

**Example 1:**   For the points $(2, 0, -4)$ and $(-1, 4, 6)$, find
(a)  the distance between the two points, and
(b)  the midpoint of the line segment joining them.

The **standard equation of a sphere** whose center is $(x_0, y_0, z_0)$ and

whose radius is $r$ is _____.

**Example 2:**   Find the center and radius of the sphere whose
equation is $x^2 + y^2 + z^2 - 4x + 2y + 8z + 17 = 0$.

**II.  Vectors in Space** (Pages 777–779)

In space, vectors are denoted by ordered triples of the form

_____.

The **zero vector in space** is denoted by _____.

> **What you should learn**
> How to analyze vectors
> in space

© 2011 Cengage Learning. All Rights Reserved. May not be scanned, copied or duplicated, or posted to a publicly accessible website, in whole or in part.

If **v** is represented by the directed line segment from $P(p_1, p_2, p_3)$ to $Q(q_1, q_2, q_3)$, the **component form** of **v** is given by

_____

_____

_____ .

Two vectors are equal if and only if _____

_____ .

The length of $\mathbf{u} = \langle u_1, u_2, u_3 \rangle$ is:

$$\| \mathbf{u} \| = \sqrt{\underline{\hspace{4cm}}}$$

A unit vector **u** in the direction of **v** is _____ .

The sum of $\mathbf{u} = \langle u_1, u_2, u_3 \rangle$ and $\mathbf{v} = \langle v_1, v_2, v_3 \rangle$ is

$\mathbf{u} + \mathbf{v} = $ _____ .

The scalar multiple of the real number $c$ and $\mathbf{u} = \langle u_1, u_2, u_3 \rangle$ is

$c\mathbf{u} = $ _____ .

**Example 3:**    Determine whether the vectors $\langle 6, 1, -3 \rangle$ and $\langle -2, -1/3, 1 \rangle$ are parallel.

To use vectors to determine whether three points $P$, $Q$, and $R$ in space are collinear, _____

_____ .

## III. Application  (Page 779)

Describe a real-life application of vectors in space.

> **What you should learn**
> How to use three-dimensional vectors to solve real-life problems

© 2011 Cengage Learning. All Rights Reserved. May not be scanned, copied or duplicated, or posted to a publicly accessible website, in whole or in part.

**Additional notes**

**Homework Assignment**

Page(s)

Exercises

© 2011 Cengage Learning. All Rights Reserved. May not be scanned, copied or duplicated, or posted to a publicly accessible website, in whole or in part.

## Section 11.3    The Dot Product of Two Vectors

**Objective:**      In this lesson you learned how to find the dot product of
two vectors in the plane or in space.

| Course Number |
| Instructor |
| Date |

---

**Important Vocabulary**          Define each term or concept.

**Angle between two nonzero vectors**

**Orthogonal**

---

### I. The Dot Product  (Pages 783–784)

The **dot product** of $\mathbf{u} = \langle u_1, u_2 \rangle$ and $\mathbf{v} = \langle v_1, v_2 \rangle$ is

_____ .

> *What you should learn*
> How to use properties of
> the dot product of two
> vectors

The dot product of $\mathbf{u} = \langle u_1, u_2, u_3 \rangle$ and $\mathbf{v} = \langle v_1, v_2, v_3 \rangle$ is

$\mathbf{u} \bullet \mathbf{v} =$ _____ .

The dot product of two vectors yields a _____ .

Let $\mathbf{u}$, $\mathbf{v}$, and $\mathbf{w}$ be vectors in the plane or in space and let $c$ be a
scalar. Complete the following properties of the dot product:

1. $\mathbf{u} \bullet \mathbf{v} =$ _____

2. $\mathbf{0} \bullet \mathbf{v} =$ _____

3. $\mathbf{u} \bullet (\mathbf{v} + \mathbf{w}) =$ _____

4. $\mathbf{v} \bullet \mathbf{v} =$ _____

5. $c(\mathbf{u} \bullet \mathbf{v}) =$ _____ = _____

**Example 1:**   Find the dot product: $\langle 5, -4 \rangle \bullet \langle 9, -2 \rangle$.

**Example 2:**   Find the dot product of the vectors $\langle -1, 4, -2 \rangle$
and $\langle 0, -1, 5 \rangle$.

### II. Angle Between Two Vectors  (Pages 784–785)

If $\theta$ is the angle between two nonzero vectors $\mathbf{u}$ and $\mathbf{v}$, then $\theta$ can

be determined from _____ .

> *What you should learn*
> How to find the angle
> between two vectors
> using the dot product

© 2011 Cengage Learning. All Rights Reserved. May not be scanned, copied or duplicated, or posted to a publicly accessible website, in whole or in part.

**Example 3:**  Find the angle between $\mathbf{v} = \langle 5, -4 \rangle$ and
              $\mathbf{w} = \langle 9, -2 \rangle$.




An alternative way to calculate the dot product between two

vectors $\mathbf{u}$ and $\mathbf{v}$, given the angle $\theta$ between them, is

_____.


Two vectors $\mathbf{u}$ and $\mathbf{v}$ are orthogonal if _____.

Two nonzero vectors are orthogonal if and only if _____

_____.


**Example 4:**  Are the vectors $\mathbf{u} = \langle 1, -4 \rangle$ and $\mathbf{v} = \langle 6, 2 \rangle$
              orthogonal?




**III. Direction Cosines**  (Page 786)

For a vector in the plane, it is convenient to measure direction in

terms of the angle, measured counterclockwise, from _____

_____. In space it is more

convenient to measure direction in terms of _____

_____

_____. The angles $\alpha$, $\beta$, and $\gamma$ are the _____

_____, and $\cos \alpha$, $\cos \beta$, and $\cos \gamma$ are the _____

_____.


The measure of $\alpha$, the angle between $\mathbf{v}$ and $\mathbf{i}$, can be found from

_____. The measure of $\beta$, the angle

between $\mathbf{v}$ and $\mathbf{j}$, can be found from _____.

The measure of $\gamma$, the angle between $\mathbf{v}$ and $\mathbf{k}$, can be found from

_____.



Any nonzero vector $\mathbf{v}$ in space has the normalized form $\dfrac{\mathbf{v}}{\|\mathbf{v}\|} =$

_____.

> **What you should learn**
> How to find the direction
> cosines of a vector in
> space

© 2011 Cengage Learning. All Rights Reserved. May not be scanned, copied or duplicated, or posted to a publicly accessible website, in whole or in part.

The sum of the squares of the directions cosines

$$\cos^2\alpha + \cos^2\beta + \cos^2\gamma = \underline{\hspace{3cm}}.$$

## IV.  Projections and Vector Components  (Pages 787–788)

Let **u** and **v** be nonzero vectors. Moreover, let $\mathbf{u} = \mathbf{w}_1 + \mathbf{w}_2$,
where $\mathbf{w}_1$ is parallel to **v**, and $\mathbf{w}_2$ is orthogonal to **v**. The vectors
$\mathbf{w}_1$ and $\mathbf{w}_2$ are called _____.
The vector $\mathbf{w}_1$ is called the **projection of u onto v** and is denoted
by _____. The vector $\mathbf{w}_2$ is given by
_____ , and is called the _____
_____.

> *What you should learn*
> How to find the
> projection of a vector
> onto another vector

Let **u** and **v** be nonzero vectors. The projection of **u** onto **v** is
given by $\text{proj}_{\mathbf{v}}\,\mathbf{u} = \underline{\hspace{5cm}}.$

## V.  Work  (Page 789)

The **work** $W$ done by a constant force **F** as its point of
application moves along the vector $\overrightarrow{PQ}$ is given by either of the
following:

1.

2.

> *What you should learn*
> How to use vectors to
> find the work done by a
> constant force

© 2011 Cengage Learning. All Rights Reserved. May not be scanned, copied or duplicated, or posted to a publicly accessible website, in whole or in part.

**Additional notes**

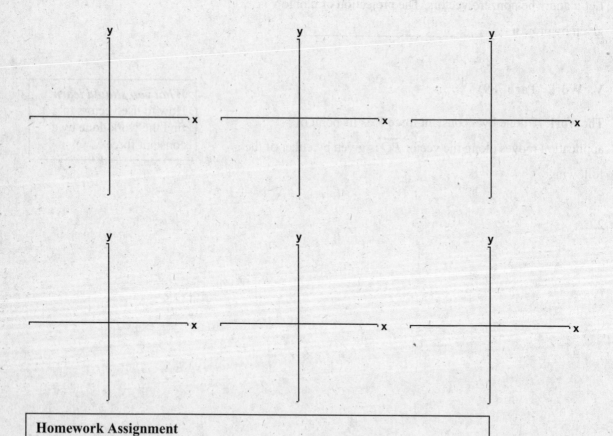

**Homework Assignment**

Page(s)

Exercises

© 2011 Cengage Learning. All Rights Reserved. May not be scanned, copied or duplicated, or posted to a publicly accessible website, in whole or in part.

## Section 11.4   The Cross Product of Two Vectors in Space

Course Number

Instructor

Date

**Objective:**   In this lesson you learned how to find the cross product of two vectors in space.

### I. The Cross Product  (Pages 792–796)

A vector in space that is orthogonal to two given vectors is called

their _____.

*What you should learn*
How to find the cross product of two vectors in space

Let $\mathbf{u} = u_1\mathbf{i} + u_2\mathbf{j} + u_3\mathbf{k}$ and $\mathbf{v} = v_1\mathbf{i} + v_2\mathbf{j} + v_3\mathbf{k}$ be two vectors in

space. The cross product of $\mathbf{u}$ and $\mathbf{v}$ is the vector

$\mathbf{u} \times \mathbf{v} = $ _____

Describe a convenient way to remember the formula for the
cross product.

**Example 1:**   Given $\mathbf{u} = -2\mathbf{i} + 3\mathbf{j} - 3\mathbf{k}$ and $\mathbf{v} = \mathbf{i} - 2\mathbf{j} + \mathbf{k}$, find
the cross product $\mathbf{u} \times \mathbf{v}$.

Let $\mathbf{u}$, $\mathbf{v}$, and $\mathbf{w}$ be vectors in space and let $c$ be a scalar.
Complete the following properties of the cross product:

1.  $\mathbf{u} \times \mathbf{v} = $ _____

2.  $\mathbf{u} \times (\mathbf{v} + \mathbf{w}) = $ _____

3.  $c(\mathbf{u} \times \mathbf{v}) = $ _____

4.  $\mathbf{u} \times \mathbf{0} = $ _____

5.  $\mathbf{u} \times \mathbf{u} = $ _____

6.  $\mathbf{u} \bullet (\mathbf{v} \times \mathbf{w}) = $ _____

Complete the following geometric properties of the cross
product, given $\mathbf{u}$ and $\mathbf{v}$ are nonzero vectors in space and $\theta$ is the
angle between $\mathbf{u}$ and $\mathbf{v}$.

1.  $\mathbf{u} \times \mathbf{v}$ is orthogonal to _____.

© 2011 Cengage Learning. All Rights Reserved. May not be scanned, copied or duplicated, or posted to a publicly accessible website, in whole or in part.

2. $\| \mathbf{u} \times \mathbf{v} \| = $ _____.

3. $\mathbf{u} \times \mathbf{v} = \mathbf{0}$ if and only if _____

_____.

4. $\| \mathbf{u} \times \mathbf{v} \| = $ area of the parallelogram having _____

_____.

## II. The Triple Scalar Product (Pages 796–797)

For vectors $\mathbf{u}$, $\mathbf{v}$, and $\mathbf{w}$ in space, the dot product of $\mathbf{u}$ and $\mathbf{v} \times \mathbf{w}$ is called the _____ of $\mathbf{u}$, $\mathbf{v}$, and $\mathbf{w}$, and is found as

> **What you should learn**
> How to use the triple scalar product of three vectors in space

$$\mathbf{u} \bullet (\mathbf{v} \times \mathbf{w}) = \left| \begin{array}{cc} & \\ & \\ & \end{array} \right|$$

The volume $V$ of a parallelepiped with vectors $\mathbf{u}$, $\mathbf{v}$, and $\mathbf{w}$ as adjacent edges is _____.

**Example 2:**  Find the volume of the parallelepiped having
$\mathbf{u} = 2\mathbf{i} + \mathbf{j} - 3\mathbf{k}$, $\mathbf{v} = \mathbf{i} - 2\mathbf{j} + 3\mathbf{k}$, and $\mathbf{w} = 4\mathbf{i} - 3\mathbf{k}$ as
adjacent edges.

---

**Homework Assignment**

Page(s)

Exercises

---

© 2011 Cengage Learning. All Rights Reserved. May not be scanned, copied or duplicated, or posted to a publicly accessible website, in whole or in part.

## Section 11.5    Lines and Planes in Space

**Objective:**    In this lesson you learned how to find equations of lines and planes in space, and how to sketch their graphs.

Course Number

Instructor

Date

### I.  Lines in Space  (Pages 800–801)

Consider the line $L$ through the point $P(x_1, y_1, z_1)$ and parallel to the vector $\mathbf{v} = \langle a, b, c \rangle$. The vector $\mathbf{v}$ is _____ for the line $L$, and $a$, $b$, and $c$ are _____.

One way of describing the line $L$ is _____

_____

_____.

A line $L$ parallel to the vector $\mathbf{v} = \langle a, b, c \rangle$ and passing through the point $P = (x_1, y_1, z_1)$ is represented by the following **parametric equations,** where $t$ is the parameter:

_____

If the direction numbers $a$, $b$, and $c$ are all nonzero, you can eliminate the parameter $t$ to obtain the **symmetric equations** of the line:

_____

***What you should learn***
How to write a set of parametric equations for a line in space

### II.  Planes in Space  (Pages 801–803)

The plane containing the point $(x_1, y_1, z_1)$ and having normal vector $\mathbf{n} = \langle a, b, c \rangle$ can be represented by the **standard form** of the equation of a plane, which is

_____

By regrouping terms, you obtain the **general form** of the equation of a plane in space:

_____

Given the general form of the equation of a plane it is easy to find a normal vector to the plane, _____

_____.

***What you should learn***
How to write a linear equation to represent a plane in space

Larson/Edwards  **Calculus: Early Transcendental Functions 5e**  Notetaking Guide

© 2011 Cengage Learning. All Rights Reserved. May not be scanned, copied or duplicated, or posted to a publicly accessible website, in whole or in part.

Two distinct planes in three-space either are _____

or _____.

If two distinct planes intersect, you can determine the angle $\theta$ between them from the angle between their normal vectors. If vectors $\mathbf{n}_1$ and $\mathbf{n}_2$ are normal to the two intersecting planes, the angle $\theta$ between the normal vectors is equal to the angle between the two planes and is given by

_____

Consequently, two planes with normal vectors $\mathbf{n}_1$ and $\mathbf{n}_2$ are

1. _____ if $\mathbf{n}_1 \bullet \mathbf{n}_2 = 0$.

2. _____ if $\mathbf{n}_1$ is a scalar multiple of $\mathbf{n}_2$.

### III. Sketching Planes in Space (Page 804)

If a plane in space intersects one of the coordinate planes, the line of intersection is called the _____ of the given plane in the coordinate plane.

To sketch a plane in space, _____

_____

_____.

The plane with equation $3y - 2z + 1 = 0$ is parallel to

_____.

> **What you should learn**
> How to sketch the plane given by a linear equation

### IV. Distances Between Points, Planes, and Lines
(Pages 805–807)

The **distance between a plane and a point** $Q$ (not in the plane) is

_____

where $P$ is a point in the plane and $\mathbf{n}$ is normal to the plane.

> **What you should learn**
> How to find the distances between points, planes, and lines in space

---

**Homework Assignment**

Page(s)

Exercises

---

© 2011 Cengage Learning. All Rights Reserved. May not be scanned, copied or duplicated, or posted to a publicly accessible website, in whole or in part.

## Section 11.6   Surfaces in Space

**Objective:**   In this lesson you learned how to recognize and write equations for cylindrical and quadric surfaces, and surfaces of revolution.

Course Number

Instructor

Date

### I.  Cylindrical Surfaces  (Pages 812–813)

Let $C$ be a curve in a plane and let $L$ be a line not in a parallel plane. The set of all lines parallel to $L$ and intersecting $C$ is called a _____. $C$ is called the _____ _____ of the cylinder, and the parallel lines are called _____.

The equation of a cylinder whose rulings are parallel to one of the coordinate axes contains only _____ _____.

*What you should learn*
How to recognize and write equations for cylindrical surfaces

### II.  Quadric Surfaces  (Pages 813–817)

Quadric surfaces are _____ _____.

The equation of a **quadric surface** in space is _____ _____. The general form of the equation is _____ _____. There are six basic types of quadric surfaces: _____ _____ _____.

The intersection of a surface with a plane is called _____ _____. To visualize a surface in space, it is helpful to _____ _____. The traces of quadric surfaces are _____.

*What you should learn*
How to recognize and write equations for quadric surfaces

© 2011 Cengage Learning. All Rights Reserved. May not be scanned, copied or duplicated, or posted to a publicly accessible website, in whole or in part.

To classify a quadric surface, _____

_____

_____

_____. For a

quadric surface not centered at the origin, you can form the

standard equation by _____.

**Example 1:**  Classify and name the center of the surface given
by $4x^2 + 36y^2 - 9z^2 + 8x - 144y + 18z + 139 = 0$.

**III.  Surfaces of Revolution**  (Page 818–819)

Consider the graph of the radius function $y = r(z)$ in the $yz$-

plane. If this graph is revolved about the $z$-axis, it forms a

_____. The trace of the

surface in the plane $z = z_0$ is a circle whose radius is $r(z_0)$ and

whose equation is _____.

| *What you should learn* |
| How to recognize and write equations for surfaces of revolution |

If the graph of a radius function r is revolved about one of the
coordinate axes, the equation of the resulting surface of
revolution has one of the following forms.

1.  Revolved about the _____ :  $y^2 + z^2 = [r(x)]^2$

2.  Revolved about the _____ :  $x^2 + z^2 = [r(y)]^2$

3.  Revolved about the _____ :  $x^2 + y^2 = [r(z)]^2$

**Homework Assignment**

Page(s)

Exercises

© 2011 Cengage Learning. All Rights Reserved. May not be scanned, copied or duplicated, or posted to a publicly accessible website, in whole or in part.

# Section 11.7    Cylindrical and Spherical Coordinates

**Objective:**    In this lesson you learned how to use cylindrical or
spherical coordinates to represent surfaces in space.

| Course Number |
| Instructor |
| Date |

**I. Cylindrical Coordinates**  (Pages 822–824)

***What you should learn***
How to use cylindrical
coordinates to represent
surfaces in space

The **cylindrical coordinate system** is an extension of _____
_____.

In a **cylindrical coordinate system,** a point $P$ in space is

represented by an ordered triple _____. $(r, \theta)$

is a polar representation of _____

_____. $z$ is the directed distance from _____

_____.

To convert from rectangular to cylindrical coordinates, or vice
versa, use the following conversion guidelines for polar
coordinates.

**Cylindrical to rectangular:**

_____    _____    _____

**Rectangular to cylindrical:**

_____    _____    _____

The point $(0, 0, 0)$ is called the _____. Because

the representation of a point in the polar coordinate system is not

unique, it follows that _____

_____.

**Example 1:**    Convert the point $(r, \theta, z) = \left(2, \dfrac{\pi}{2}, 5\right)$ to

rectangular coordinates.

Cylindrical coordinates are especially convenient for

representing _____

_____.

© 2011 Cengage Learning. All Rights Reserved. May not be scanned, copied or duplicated, or posted to a publicly accessible website, in whole or in part.

Give an example of a cylindrical coordinate equation for a
vertical plane containing the $z$-axis. _____

Give an example of a cylindrical coordinate equation for a
horizontal plane. _____

## II. Spherical Coordinates  (Pages 825–826)

In a spherical coordinate system, a point $P$ in space is
represented by an ordered triple _____.

1.  $\rho$ is the distance between _____.

2.  $\theta$ is the same angle used in _____

_____.

3.  $\phi$ is the angle between _____

_____.

> *What you should learn*
> How to use spherical
> coordinates to represent
> surfaces in space

**To convert from spherical to rectangular coordinates, use:**

_____    _____    _____

**To convert from rectangular to spherical coordinates, use:**

_____    _____    _____

**To convert from spherical to cylindrical coordinates ($r \geq 0$), use:**

_____    _____    _____

**To convert from cylindrical to spherical coordinates ($r \geq 0$), use:**

_____    _____    _____

---

**Homework Assignment**

Page(s)

Exercises

---

© 2011 Cengage Learning. All Rights Reserved. May not be scanned, copied or duplicated, or posted to a publicly accessible website, in whole or in part.

# Chapter 12     Vector-Valued Functions

## Section 12.1     Vector-Valued Functions

| Course Number |
| Instructor |
| Date |

**Objective:** In this lesson you learned how to analyze and sketch a space curve represented by a vector-valued function and how to apply the concepts of limits and continuity to vector-valued functions.

### I. Space Curves and Vector-Valued Functions
(Pages 834–836)

> **What you should learn**
> How to analyze and sketch a space curve given by a vector-valued function

A **space curve** $C$ is the set of all ordered triples _____ _____ together with their defining parametric equations _____ _____ where $f$, $g$, and $h$ are continuous functions of $t$ on an interval $I$.

A function of the form $\mathbf{r}(t) = f(t)\mathbf{i} + g(t)\mathbf{j}$ in a plane or

$\mathbf{r}(t) = f(t)\mathbf{i} + g(t)\mathbf{j} + h(t)\mathbf{k}$ in space is a _____

_____, where the **component functions** $f$, $g$, and $h$ are real-valued functions of the parameter $t$. Vector-valued functions are sometimes denoted as _____ or ____

_____.

Vector-valued functions serve dual roles in the representation of curves. By letting the parameter $t$ represent time, you can use a vector-valued function to represent _____

_____. Or, in the more general case, you can use a vector-valued function to _____. In either case, the terminal point of the position vector $\mathbf{r}(t)$ coincides with _____

_____. The arrowhead on the curve indicates the curve's _____ by pointing in the direction of increasing values of $t$.

Unless stated otherwise, the **domain** of a vector-valued function $\mathbf{r}$ is considered to be _____

_____.

© 2011 Cengage Learning. All Rights Reserved. May not be scanned, copied or duplicated, or posted to a publicly accessible website, in whole or in part.

## II.  Limits and Continuity  (Pages 837–838)

**What you should learn**
How to extend the
concepts of limits and
continuity to vector-
valued functions

### Definition of the Limit of a Vector-Valued Function

1.  If **r** is a vector-valued function such that $\mathbf{r}(t) = f(t)\mathbf{i} + g(t)\mathbf{j}$,

    then _____,

    provided $f$ and $g$ have limits as $t \to a$.

2.  If **r** is a vector-valued function in space such that
    $\mathbf{r}(t) = f(t)\mathbf{i} + g(t)\mathbf{j} + h(t)\mathbf{k}$ , then

    _____,

    provided $f$, $g$, and $h$ have limits as $t \to a$.

If $\mathbf{r}(t)$ approaches the vector **L** as $t \to a$, the length of the vector
$\mathbf{r}(t) - \mathbf{L}$ approaches _____.

A vector-valued function **r** is **continuous at the point** given by
$t = a$ if _____
_____. A vector-valued function **r** is
**continuous on an interval** $I$ if _____
_____.

---

**Homework Assignment**

Page(s)

Exercises

© 2011 Cengage Learning. All Rights Reserved. May not be scanned, copied or duplicated, or posted to a publicly accessible website, in whole or in part.

## Section 12.2   Differentiation and Integration of Vector-Valued Functions

**Objective:**   In this lesson you learned how to differentiate and integrate vector-valued functions.

**I. Differentiation of Vector-Valued Functions**
(Pages 842–845)

**What you should learn**
How to differentiate a
vector-valued function

The **derivative of a vector-valued function r** is defined by

_____ for all $t$ for

which the

limit exists. If $\mathbf{r}'(t)$ exists, then $\mathbf{r}$ is _____.

If $\mathbf{r}'(t)$ exists for all $t$ in an open interval $I$, then $\mathbf{r}$ is

_____.

Differentiability of vector-valued functions can be extended to

closed intervals by _____.

If $\mathbf{r}(t) = f(t)\mathbf{i} + g(t)\mathbf{j}$, where $f$ and $g$ are differentiable functions of

$t$, then _____.

If $\mathbf{r}(t) = f(t)\mathbf{i} + g(t)\mathbf{j} + h(t)\mathbf{k}$, where $f$, $g$, and $h$ are differentiable

functions of $t$, then _____.

**Example 1:**   Find $\mathbf{r}'(t)$ for the vector-valued function given by
$\mathbf{r}(t) = (1 - t^2)\mathbf{i} + 5\mathbf{j} + \ln t\mathbf{k}$.

The parameterization of the curve represented by the vector-

valued function $\mathbf{r}(t) = f(t)\mathbf{i} + g(t)\mathbf{j} + h(t)\mathbf{k}$ is **smooth on an open**

**interval** $I$ if _____

_____.

Let $\mathbf{r}$ and $\mathbf{u}$ be differentiable vector-valued functions of $t$, let $w$
be a differentiable real-valued function of $t$, and let $c$ be a scalar.

1.  $D_t[c\mathbf{r}(t)] = $ _____.

2.  $D_t[\mathbf{r}(t) \pm \mathbf{u}(t)] = $ _____.

© 2011 Cengage Learning. All Rights Reserved. May not be scanned, copied or duplicated, or posted to a publicly accessible website, in whole or in part.

3. $D_t[w(t)\mathbf{r}(t)] =$ _____.

4. $D_t[\mathbf{r}(t) \cdot \mathbf{u}(t)] =$ _____.

5. $D_t[\mathbf{r}(t) \times \mathbf{u}(t)] =$ _____.

6. $D_t[\mathbf{r}(w(t))] =$ _____.

7. If $\mathbf{r}(t) \cdot \mathbf{r}(t) = c$, then _____.

## II. Integration of Vector-Valued Functions (Pages 846–847)

If $\mathbf{r}(t) = f(t)\mathbf{i} + g(t)\mathbf{j}$, where $f$ and $g$ are continuous on $[a, b]$,

then the _____ is

$$\int \mathbf{r}(t)\, dt = \left[\int f(t)\, dt\right]\mathbf{i} + \left[\int g(t)\, dt\right]\mathbf{j} \text{ and its } \textbf{definite integral}$$

over the interval _____ is

$$\int_a^b \mathbf{r}(t)\, dt = \left[\int_a^b f(t)\, dt\right]\mathbf{i} + \left[\int_a^b g(t)\, dt\right]\mathbf{j}.$$

If $\mathbf{r}(t) = f(t)\mathbf{i} + g(t)\mathbf{j} + h(t)\mathbf{k}$, where $f$, $g$, and $h$ are continuous

on $[a, b]$, then the _____ is

$$\int \mathbf{r}(t)\, dt = \left[\int f(t)\, dt\right]\mathbf{i} + \left[\int g(t)\, dt\right]\mathbf{j} + \left[\int h(t)\, dt\right]\mathbf{k} \text{ and its}$$

**definite integral** over the interval _____ is

$$\int_a^b \mathbf{r}(t)\, dt = \left[\int_a^b f(t)\, dt\right]\mathbf{i} + \left[\int_a^b g(t)\, dt\right]\mathbf{j} + \left[\int_a^b h(t)\, dt\right]\mathbf{k}.$$

The antiderivative of a vector-valued function is a family of

vector-valued functions all differing by _____

_____.

*What you should learn*
How to integrate a
vector-valued function

---

**Homework Assignment**

Page(s)

Exercises

© 2011 Cengage Learning. All Rights Reserved. May not be scanned, copied or duplicated, or posted to a publicly accessible website, in whole or in part.

## Section 12.3    Velocity and Acceleration

**Objective:**    In this lesson you learned how to describe the velocity and acceleration associated with a vector-valued function and how to use a vector-valued function to analyze projectile motion.

Course Number

Instructor

Date

### I. Velocity and Acceleration  (Pages 850–853)

If $x$ and $y$ are twice-differentiable function of $t$, and $\mathbf{r}$ is a vector-valued function given by $\mathbf{r}(t) = x(t)\mathbf{i} + y(t)\mathbf{j}$, then the velocity vector, acceleration vector, and speed at time $t$ are as follows.

1. **Velocity** = $\mathbf{v}(t)$ = _____.

2. **Acceleration** = $\mathbf{a}(t)$ = _____.

3. **Speed** = $\|\mathbf{v}(t)\|$ = _____ .

*What you should learn*
How to describe the velocity and acceleration associated with a vector-valued function

List the corresponding definitions for velocity, acceleration, and speed along a space curve given by $\mathbf{r}(t) = x(t)\mathbf{i} + y(t)\mathbf{j} + z(t)\mathbf{k}$.

**Example 1:**    Find the velocity vector and acceleration vector of a particle that moves along the plane curve $C$ given by $\mathbf{r}(t) = \cos t\,\mathbf{i} - 2t\mathbf{j}$ .

### II. Projectile Motion  (Pages 854–855)

Neglecting air resistance, the path of a projectile launched from an initial height $h$ with initial speed $v_0$ and angle of elevation $\theta$ is described by the vector function

*What you should learn*
How to use a vector-valued function to analyze projectile motion

_____

where $g$ is the acceleration due to gravity.

© 2011 Cengage Learning. All Rights Reserved. May not be scanned, copied or duplicated, or posted to a publicly accessible website, in whole or in part.

**Additional notes**

**Homework Assignment**

Page(s)

Exercises

© 2011 Cengage Learning. All Rights Reserved. May not be scanned, copied or duplicated, or posted to a publicly accessible website, in whole or in part.

## Section 12.4   Tangent Vectors and Normal Vectors

**Objective:**      In this lesson you learned how to find tangent vectors and normal vectors.

Course Number

Instructor

Date

**I.  Tangent Vectors and Normal Vectors**  (Pages 859–862)

Let $C$ be a smooth curve represented by **r** on an open interval $I$. The **unit tangent vector** $\mathbf{T}(t)$ at $t$ is defined to be _____ _____ .

Recall that a curve is smooth on an interval if _____ _____ . So, "smoothness" is sufficient to guarantee that _____ _____ .

The tangent line to a curve at a point is _____ _____ .

Let $C$ be a smooth curve represented by **r** on an open interval $I$. If $\mathbf{T}'(t) \neq \mathbf{0}$, then the **principal unit normal vector** at $t$ is defined to be _____ .

**What you should learn**
How to find a unit tangent vector at a point on a space curve

**II.  Tangential and Normal Components of Acceleration**
(Pages 862–865)

For an object traveling at a constant speed, the velocity and acceleration vectors _____ . For an object traveling at a variable speed, the velocity and acceleration vectors _____ .

If **r**$(t)$ is the position vector for a smooth curve $C$ and $\mathbf{N}(t)$ exists, then the acceleration vector **a**$(t)$ lies _____ _____ .

If **r**$(t)$ is the position vector for a smooth curve $C$ [for which $\mathbf{N}(t)$ exists], then the **tangential component of acceleration** $a_{\mathrm{T}}$ and the **normal component of acceleration** $a_{\mathrm{N}}$ are as follows.

**What you should learn**
How to find the tangential and normal components of acceleration

© 2011 Cengage Learning. All Rights Reserved. May not be scanned, copied or duplicated, or posted to a publicly accessible website, in whole or in part.

$a_T =$

_____

$a_N =$

_____

Note that $a_N \geq 0$. The normal component of acceleration is also

called the _____.


**Additional notes**

**Homework Assignment**

Page(s)

Exercises

© 2011 Cengage Learning. All Rights Reserved. May not be scanned, copied or duplicated, or posted to a publicly accessible website, in whole or in part.

## Section 12.5   Arc Length and Curvature

**Objective:**   In this lesson you learned how to find the arc length and
curvature of a curve.

Course Number

Instructor

Date

**I. Arc Length and Curvature**  (Pages 869–870)

If $C$ is a smooth curve given by $\mathbf{r}(t) = x(t)\mathbf{i} + y(t)\mathbf{j} + z(t)\mathbf{k}$, on an
interval $[a, b]$, then the arc length of $C$ on the interval is

_____ .

**What you should learn**
How to find a unit
tangent vector at a point
on a space curve

**Example 1:**   Find the arc length of the curve given by
$\mathbf{r}(t) = \sin t\mathbf{i} - 2t\mathbf{j} + t^2\mathbf{k}$, from $t = 0$ to $t = 4$.

**II. Arc Length Parameter**  (Pages 870–871)

Let $C$ be a smooth curve given by $\mathbf{r}(t)$ defined on the closed
interval $[a, b]$. For $a \le t \le b$, the arc length function is given by

_____ .

**What you should learn**
How to find the
tangential and normal
components of
acceleration

The arc length $s$ is called the _____ .

If $C$ is a smooth curve given by $\mathbf{r}(s) = x(s)\mathbf{i} + y(s)\mathbf{j}$ or

$\mathbf{r}(s) = x(s)\mathbf{i} + y(s)\mathbf{j} + z(s)\mathbf{k}$ where $s$ is the arc length parameter,

then _____ . Moreover, if $t$ is *any*

parameter for the vector-valued function $\mathbf{r}$ such that $\|\mathbf{r}'(t)\| = 1$,

then $t$ _____ .

**III. Curvature**  (Pages 872–875)

**Curvature** is the measure of _____

_____ .

**What you should learn**
How to find the
tangential and normal
components of
acceleration

© 2011 Cengage Learning. All Rights Reserved. May not be scanned, copied or duplicated, or posted to a publicly accessible website, in whole or in part.

Let $C$ be a smooth curve (in the plane or in space) given by $\mathbf{r}(s)$, where $s$ is the arc length parameter. The **curvature** $K$ at $s$ is

given by _____ .


Describe the curvature of a circle.


_____ .


If $C$ is a smooth curve given by $\mathbf{r}(t)$, then two additional formulas for finding the curvature $K$ of $C$ at $t$ are

$K = $ _____ , or


$K = $ _____ .


If $C$ is the graph of a twice-differentiable function given by $y = f(x)$, then the curvature $K$ at the point $(x, y)$ is given by

$K = $ _____ .


Let $C$ be a curve with curvature $K$ at point $P$. The circle passing through point $P$ with radius $r = 1/K$ is called the **circle of curvature** if _____

_____
_____ . The radius is called the _____
_____ at $P$, and the center of the circle is called
the _____ .


If $\mathbf{r}(t)$ is the position vector for a smooth curve $C$, then the acceleration vector is given by

_____ , where $K$ is the curvature

of $C$ and $ds/dt$ is the speed.

© 2011 Cengage Learning. All Rights Reserved. May not be scanned, copied or duplicated, or posted to a publicly accessible website, in whole or in part.

**IV.  Application**  (Pages 876–877)

A moving object with mass $m$ is in contact with a stationary object. The total force required to produce an acceleration **a** along a given path is

*What you should learn*
How to find the tangential and normal components of acceleration

The portion of this total force that is supplied by the stationary

object is called the _____.


**Additional notes**

© 2011 Cengage Learning. All Rights Reserved. May not be scanned, copied or duplicated, or posted to a publicly accessible website, in whole or in part.

**Additional notes**

**Homework Assignment**

Page(s)

Exercises

© 2011 Cengage Learning. All Rights Reserved. May not be scanned, copied or duplicated, or posted to a publicly accessible website, in whole or in part.

# Chapter 13    Functions of Several Variables

Course Number

Instructor

Date

**Section 13.1**      **Introduction to Functions of Several Variables**

**Objective:** In this lesson you learned how to sketch a graph, level curves, and level surfaces.

---

**Important Vocabulary**      Define each term or concept.

**Function of two variables**

**Domain of a function of two variables**

**Range of a function of two variables**

---

## I. Functions of Several Variables (Pages 886–887)

For the function given by $z = f(x, y)$, $x$ and $y$ are called the

_____ and $z$ is called the

_____ of $f$.

> *What you should learn*
> How to understand the notation for a function of several variables

**Example 1:** For $f(x, y) = \sqrt{100 - 2x^2 - 6y}$, evaluate $f(3, 3)$.

**Example 2:** For $f(x, y, z) = 2x + 5y^2 - z^3$, evaluate $f(4, 3, 2)$.

## II. The Graph of a Function of Two Variables (Page 888)

The **graph** of a function $f$ of two variables is _____

_____

_____.

> *What you should learn*
> How to sketch the graph of a function of two variables

© 2011 Cengage Learning. All Rights Reserved. May not be scanned, copied or duplicated, or posted to a publicly accessible website, in whole or in part.

The graph of $z = f(x, y)$ is a surface whose projection onto the

$xy$-plane is _____. To each

point $(x, y)$ in $D$ there corresponds _____

_____, and conversely, to each point $(x, y, z)$ on

the surface there corresponds _____.


To sketch a surface in space by hand, it helps to use _____

_____.


### III.  Level Curves  (Pages 889–891)

A second way to visualize a function of two variables is to use a

_____ in which the scalar $z = f(x, y)$ is

assigned to the point $(x, y)$. A scalar field can be characterized by

_____ or _____ along

which the value of $f(x, y)$ is _____.

Name a few applications of level curves.

**What you should learn**
How to sketch level
curves for a function of
two variables

A contour map depicts _____

_____. Much

space between level curves indicates that _____

_____, whereas little space indicates _____

_____.


What is the **Cobb-Douglas production function**?

© 2011 Cengage Learning. All Rights Reserved. May not be scanned, copied or duplicated, or posted to a publicly accessible website, in whole or in part.

Let $x$ measure the number of units of labor and let $y$ measure the number of units of capital. Then the number of units produced is modeled by the function

_____

**Example 3:**   A manufacturer estimates that its production
(measured in units of a product) can be modeled
by $f(x, y) = 400x^{0.3}y^{0.7}$, where the labor $x$ is
measured in person-hours and the capital $y$ is
measured in thousands of dollars. What is the
production level when $x = 500$ and $y = 200$?

**IV.  Level Surfaces**  (Pages 891–892)

The concept of a level curve can be extended by one dimension

to define a _____. If $f$ is a

function of three variables and $c$ is a constant, the graph of the

equation $f(x, y, z) = c$ is _____

_____.

> *What you should learn*
> How to sketch level
> curves for a function of
> three variables

**V.  Computer Graphics**  (Pages 892–893)

The problem of sketching the graph of a surface can be

simplified by _____.

> *What you should learn*
> How to use computer
> graphs to graph a
> function of two variables

© 2011 Cengage Learning. All Rights Reserved. May not be scanned, copied or duplicated, or posted to a publicly accessible website, in whole or in part.

**Additional notes**

```
+--------------------------------------------------+
| Homework Assignment                              |
|                                                  |
| Page(s)                                          |
|                                                  |
| Exercises                                        |
|                                                  |
+--------------------------------------------------+
```

© 2011 Cengage Learning. All Rights Reserved. May not be scanned, copied or duplicated, or posted to a publicly accessible website, in whole or in part.

## Section 13.2   Limits and Continuity

**Objective:**      In this lesson you learned how to find a limit and
determine continuity.

Course Number

Instructor

Date

### I.  Neighborhoods in the Plane  (Page 898)

Using the formula for the distance between two points $(x, y)$ and
$(x_0, y_0)$ in the plane, you can define the $\delta$-neighborhood about
$(x_0, y_0)$ to be _____.

When this formula $\left\{(x, y): \ \sqrt{(x-x_0)^2 + (y-y_0)^2} < \delta\right\}$

contains the less than inequality, $<$, the disk is called

_____. When it contains the less than or equal to

inequality, $\leq$, the disk is called _____.

A point $(x_0, y_0)$ in a plane region $R$ is an **interior point** of $R$ if

there exists _____

_____. If every point in $R$ is an

interior point, then R is _____. A point

$(x_0, y_0)$ is a **boundary point** of $R$ if _____

_____

_____. By definition, a region must

contain its interior points, but it need not contain _____

_____. If a region contains all its boundary

points, the region is _____. A region that contains

some but not all of its boundary points is _____

_____.

*What you should learn*
How to understand the
definition of a
neighborhood in the
plane

### II.  Limit of a Function of Two Variables  (Pages 899–901)

Let $f$ be a function of two variables defined, except possibly at
$(x_0, y_0)$, on an open disk centered at $(x_0, y_0)$, and let $L$ be a real

number. Then $\lim\limits_{(x,y)\to(x_0,y_0)} f(x,y) = $ _____ if for each

*What you should learn*
How to understand and
use the definition of the
limit of a function of two
variables

© 2011 Cengage Learning. All Rights Reserved. May not be scanned, copied or duplicated, or posted to a publicly accessible website, in whole or in part.

$\varepsilon > 0$ there corresponds _____ such that

$|f(x,y)-L|<\varepsilon$ whenever $0<\sqrt{(x-x_0)^2+(y-y_0)^2}<\delta$.

For a function of two variables, the statement $(x,y)\to(x_0,y_0)$

means _____

_____. If the

value of $\lim\limits_{(x,y)\to(x_0,y_0)} f(x,y)$ is not the same for all possible

approaches, or **paths,** to $(x_0, y_0)$, _____

_____.

**Example 1:**   Evaluate   $\lim\limits_{(x,y)\to(4,-1)} \dfrac{x^2+16y}{3x-4y}$.

**III.  Continuity of a Function of Two Variables**  (Pages 902–903)

A function $f$ of two variables is **continuous at a point $(x_0, y_0)$** in

an open region $R$ if _____

_____.

The function $f$ is _____ if

it is continuous at every point in $R$.

Discuss the difference between **removable** and **nonremovable** discontinuities.

> **What you should learn**
> How to extend the concept of continuity to a function of two variables

If $k$ is a real number and $f$ and $g$ are continuous at $(x_0, y_0)$, then the following functions are continuous at $(x_0, y_0)$.

1.

2.

3.

4.

© 2011 Cengage Learning. All Rights Reserved. May not be scanned, copied or duplicated, or posted to a publicly accessible website, in whole or in part.

If $h$ is continuous at $(x_0, y_0)$ and $g$ is continuous at $h(x_0, y_0)$, then

the composite function given by $(g \circ h)(x, y) = g(h(x, y))$ is

_____. That is,

$$\lim_{(x,y) \to (x_0, y_0)} g(h(x, y)) = g(h(x_0, y_0)).$$

## IV.  Continuity of a Function of Three Variables  (Page 904)

A function $f$ of three variables is **continuous at a point**

$(x_0, y_0, z_0)$ in an open region $R$ if _____

_____

_____.

That is,    $\lim_{(x,y,z) \to (x_0, y_0, z_0)} f(x, y, z) = f(x_0, y_0, z_0)$ . The function $f$ is

_____ if it is

continuous at every point in $R$.

**What you should learn**
How to extend the
concept of continuity to a
function of three
variables

© 2011 Cengage Learning. All Rights Reserved. May not be scanned, copied or duplicated, or posted to a publicly accessible website, in whole or in part.

**Additional notes**

**Homework Assignment**

Page(s)

Exercises

© 2011 Cengage Learning. All Rights Reserved. May not be scanned, copied or duplicated, or posted to a publicly accessible website, in whole or in part.

Course Number

Instructor

Date

## Section 13.3   Partial Derivatives

**Objective:**   In this lesson you learned how to find and use a partial derivative.

*What you should learn*
How to find and use partial derivatives of a function of two variables

### I. Partial Derivatives of a Function of Two Variables
   (Pages 908–911)

The process of determining the rate of change of a function $f$ with respect to one of its several independent variables is called _____, and the result is referred to as the _____ of $f$ with respect to the chosen independent variable.

If $z = f(x, y)$, then the **first partial derivatives** of $f$ with respect to $x$ and $y$ are the functions $f_x$ and $f_y$, defined by

$$f_x(x, y) = \lim_{\Delta x \to 0} \underline{\hspace{4cm}}$$

$$f_y(x, y) = \lim_{\Delta y \to 0} \underline{\hspace{4cm}}$$

provided the limit exists.

This definition indicates that if $z = f(x, y)$, then to find $f_x$, you consider _____ _____. Similarly, to find $f_y$, you consider _____ _____.

List the equivalent ways of denoting the first partial derivatives of $z = f(x, y)$ with respect to $x$.

List the equivalent ways of denoting the first partial derivatives of $z = f(x, y)$ with respect to $y$.

© 2011 Cengage Learning. All Rights Reserved. May not be scanned, copied or duplicated, or posted to a publicly accessible website, in whole or in part.

The values of the first partial derivatives at the point $(a, b)$ are denoted by

$$\left.\frac{\partial z}{\partial x}\right|_{(a,b)} = \underline{\hspace{3cm}} \qquad \text{and} \qquad \left.\frac{\partial z}{\partial y}\right|_{(a,b)} = \underline{\hspace{3cm}}$$

**Example 1:** Find $\partial z/\partial x$ for the function
$$z = 20 - 2x^2 + 3xy + 5x^2 y^2.$$

For the function $z = f(x, y)$, if $y = y_0$, then $z = f(x, y_0)$

represents the curve formed by intersecting _____

_____ . On this curve, the

partial derivative $f_x(x_0, y_0)$ represents _____

_____ .

Informally, the values of $\partial f / \partial x$ and $\partial f / \partial y$ at the point

$(x_0, y_0, z_0)$ denote _____

_____ , respectively.

**Example 2:** Find the slope of the surface given by
$$z = 20 - 2x^2 + 3xy + 5x^2 y^2 \text{ at the point } (1, 1, 26) \text{ in}$$
the $y$-direction.

**II. Partial Derivatives of a Function of Three or More
Variables** (Pages 911–912)

The function $w = f(x, y, z)$ has _____ partial

derivatives, each of which is formed by _____

_____ .

The partial derivative of $w$ with respect to $x$ is denoted by

_____ . To find the partial

derivative of $w$ with respect to $x$, consider _____ to

be constant and differentiate with respect to _____ .

| What you should learn |
| --- |
| How to find and use partial derivatives of a function of three or more variables |

© 2011 Cengage Learning. All Rights Reserved. May not be scanned, copied or duplicated, or posted to a publicly accessible website, in whole or in part.

### III.  Higher-Order Partial Derivatives  (Pages 912–913)

*What you should learn*
How to find higher-order partial derivatives of a function of two or three variables

As with ordinary derivatives, it is possible to take _____ _____ partial derivatives of a function of several variables, provided such derivatives exist. Higher-order derivatives are denoted by _____

_____.

The notation $\dfrac{\partial^2 f}{\partial x \partial y}$ indicates to differentiate first with respect to _____ and then with respect to _____.

The notation $\dfrac{\partial}{\partial y}\left(\dfrac{\partial f}{\partial x}\right)$ indicates to differentiate first with respect to _____ and then with respect to _____.

The notation $f_{yx}$ indicates to differentiate first with respect to _____ and then with respect to _____.

The cases represented in the three examples of notation given above are called _____.

**Example 3:**  Find the value of $f_{xy}(2,-3)$ for the function
$$f(x, y) = 20 - 2x^2 + 3xy + 5x^2 y^2.$$

If $f$ is a function of $x$ and $y$ such that $f_{xy}$ and $f_{yx}$ are continuous on an open disk $R$, then, for every $(x, y)$ in $R$,

$f_{xy}(x, y) =$ _____.

© 2011 Cengage Learning. All Rights Reserved. May not be scanned, copied or duplicated, or posted to a publicly accessible website, in whole or in part.

**Additional notes**

**Homework Assignment**

Page(s)

Exercises

© 2011 Cengage Learning. All Rights Reserved. May not be scanned, copied or duplicated, or posted to a publicly accessible website, in whole or in part.

## Section 13.4    Differentials

**Objective:**    In this lesson you learned how to find and use a total
differential and determine differentiability.

Course Number

Instructor

Date

**I. Increments and Differentials**  (Page 918)

The **increments of $x$ and $y$** are _____, and the

**increment of $z$** is given by _____

_____.

*What you should learn*
How to understand the
concepts of increments
and differentials

If $z = f(x, y)$ and $\Delta x$ and $\Delta y$ are increments of $x$ and $y$, then the

**differentials** of the independent variables $x$ and $y$ are _____

_____, and the total differential of the

dependent variable $z$ is

_____.

**II. Differentiability**  (Page 919)

A function $f$ given by $z = f(x, y)$ is **differentiable** at $(x_0, y_0)$ if

$\Delta z$ can be written in the form

*What you should learn*
How to extend the
concept of
differentiability to a
function of two variables

_____

where both $\varepsilon_1$ and $\varepsilon_2 \to 0$ as $(\Delta x, \Delta y) \to (0, 0)$. The function $f$ is

_____ if it is

differentiable at each point in $R$.

If $f$ is a function of $x$ and $y$, where $f_x$ and $f_y$ are continuous in an

open region $R$, then $f$ is _____.

**III. Approximation by Differentials**  (Pages 920–922)

The partial derivatives $\partial z/\partial x$ and $\partial z/\partial y$ can be interpreted as

*What you should learn*
How to use a differential
as an approximation

_____.

© 2011 Cengage Learning. All Rights Reserved. May not be scanned, copied or duplicated, or posted to a publicly accessible website, in whole or in part.

This means that $dz = \dfrac{\partial z}{\partial x}\Delta x + \dfrac{\partial z}{\partial y}\Delta y$ represents _____

_____

_____. Because a plane in

space is represented by a linear equation in the variables $x$, $y$, and

$z$, the approximation of $\Delta z$ by $dz$ is called a _____

_____.

If a function of $x$ and $y$ is differentiable at $(x_0, y_0)$, then _____

_____.

Homework Assignment

Page(s)

Exercises

© 2011 Cengage Learning. All Rights Reserved. May not be scanned, copied or duplicated, or posted to a publicly accessible website, in whole or in part.

## Section 13.5  Chain Rules for Functions of Several Variables

Course Number

Instructor

Date

**Objective:**    In this lesson you learned how to use the Chain Rules and find a partial derivative implicitly.

### I. Chain Rules for Functions of Several Variables
   (Pages 925–929)

*What you should learn*
How to use the Chain
Rules for functions of
several variables

Let $w = f(x, y)$, where $f$ is a differentiable function of $x$ and $y$.

The Chain Rule for One Independent Variable states that _____

_____

_____

_____ .

**Example 1:**   Let $w = 2xy + 3xy^3$, where $x = 1 - 2t$ and $y = 2\sin t$. Find $dw/dt$.

Let $w = f(x, y)$, where $f$ is a differentiable function of $x$ and $y$.

The Chain Rule for Two Independent Variables states that _____

_____

_____

_____

_____

_____ .

### II. Implicit Partial Differentiation  (Pages 929–930)

*What you should learn*
How to find partial
derivatives implicitly

If the equation $F(x, y) = 0$ defines $y$ implicitly as a

differentiable function of $x$, then $\dfrac{dy}{dx} = $ _____ ,

$F_y(x, y) \neq 0$. If the equation $F(x, y, z) = 0$ defines $z$ implicitly as

© 2011 Cengage Learning. All Rights Reserved. May not be scanned, copied or duplicated, or posted to a publicly accessible website, in whole or in part.

a differentiable function of $x$ and $y$, then

$$\frac{\partial z}{\partial x} = \underline{\hspace{6cm}}, \text{ and}$$

$$\frac{\partial z}{\partial y} = \underline{\hspace{6cm}},$$

$F_z(x, y, z) \neq 0$.

**Example 2:** Find $dy/dx$, given $2x^2 + xy + y^2 - x - 2y = 0$.

---

**Homework Assignment**

Page(s)

Exercises

---

© 2011 Cengage Learning. All Rights Reserved. May not be scanned, copied or duplicated, or posted to a publicly accessible website, in whole or in part.

## Section 13.6   Directional Derivatives and Gradients

**Objective:**    In this lesson you learned how to find and use a
directional derivative and a gradient.

Course Number

Instructor

Date

### I.  Directional Derivative  (Pages 933–935)

Let $f$ be a function of two variables $x$ and $y$, and let

$\mathbf{u} = \cos\theta\mathbf{i} + \sin\theta\mathbf{j}$  be a unit vector. Then the _____

_____ , denoted by

$D_\mathbf{u}f$, is  $D_\mathbf{u}f(x,y) = \lim\limits_{t\to 0}\dfrac{f(x+t\cos\theta,\, y+t\sin\theta) - f(x,y)}{t}$,

provided this limit exists.

A simpler working formula for finding a directional derivative
states that if $f$ is a differentiable function of $x$ and $y$, then the
directional derivative of $f$ in the direction of the unit vector

$\mathbf{u} = \cos\theta\mathbf{i} + \sin\theta\mathbf{j}$  is _____ .

> **What you should learn**
> How to find and use
> directional derivatives of
> a function of two
> variables

### II.  The Gradient of a Function of Two Variables
   (Pages 936–937)

Let  $z = f(x,y)$  be a function of $x$ and $y$ such that  $f_x$  and  $f_y$

exist. Then the **gradient of** $f$, denoted by _____ ,

is the vector _____ .

Note that for each $(x, y)$, the gradient $\nabla f(x, y)$ is a vector in

_____ .

If $f$ is a differentiable function of $x$ and $y$, then the directional

derivative of $f$ in the direction of the unit vector $\mathbf{u}$ is

_____ .

> **What you should learn**
> How to find the gradient
> of a function of two
> variables

### III.  Applications of the Gradient  (Pages 937–940)

In many applications, you may want to know in which direction

to move so that $f(x, y)$ increases most rapidly. This direction is

> **What you should learn**
> How to use the gradient
> of a function of two
> variables in applications

© 2011 Cengage Learning. All Rights Reserved. May not be scanned, copied or duplicated, or posted to a publicly accessible website, in whole or in part.

called _____,

and it is given by the _____.

Let $f$ be differentiable at the point $(x, y)$. State three properties of the gradient at that point.

1.

2.

3.

If $f$ is differentiable at $(x_0, y_0)$ and $\nabla f(x_0, y_0) \neq \mathbf{0}$, then $\nabla f(x_0, y_0)$ is

_____.

## IV. Functions of Three Variables (Page 941)

Let $f$ be a function of $x$, $y$, and $z$, with continuous first partial derivatives. The **directional derivative of $f$** in the direction of a unit vector $\mathbf{u} = a\mathbf{i} + b\mathbf{j} + c\mathbf{k}$ is given by

_____.

The **gradient of $f$** is defined to be _____

_____.

Properties of the gradient are as follows.

1.

2.

3.

4.

> **What you should learn**
> How to find directional derivatives and gradients of functions of three variables

```
Homework Assignment
Page(s)

Exercises
```

© 2011 Cengage Learning. All Rights Reserved. May not be scanned, copied or duplicated, or posted to a publicly accessible website, in whole or in part.

## Section 13.7   Tangent Planes and Normal Lines

Course Number

Instructor

Date

**Objective:**   In this lesson you learned how to find and use a directional derivative and a gradient.

### I. Tangent Plane and Normal line to a Surface
   (Pages 945–949)

***What you should learn***
How to find equations of tangent planes and normal lines to surfaces

For a surface $S$ given by $z = f(x, y)$, you can convert to the general form by defining $F$ as $F(x, y, z) =$ _____.

Because $f(x, y) - z = 0$, you can consider $S$ to be _____

_____.

**Example 1:**   For the function given by
$F(x, y, z) = 12 - 3x^2 + y^2 - 4z^2$, describe the level surface given by $F(x, y, z) = 0$.

Let $F$ be differentiable at the point $P(x_0, y_0, z_0)$ on the surface $S$ given by $F(x, y, z) = 0$ such that $\nabla F(x_0, y_0, z_0) \neq \mathbf{0}$.

1.   The plane through $P$ that is normal to $\nabla F(x_0, y_0, z_0)$ is called

   _____.

2.   The line through $P$ having the direction of $\nabla F(x_0, y_0, z_0)$ is

   called _____.

If $F$ is differentiable at $(x_0, y_0, z_0)$, then an equation of the tangent plane to the surface given by $F(x, y, z) = 0$ at $(x_0, y_0, z_0)$ is

_____

To find the equation of the tangent plane at a point on a surface given by $z = f(x, y)$, you can define the function $F$ by $F(x, y, z) = f(x, y) - z$. Then $S$ is given by the level surface $F(x, y, z) = 0$, and an equation of the tangent plane to $S$ at the point $(x_0, y_0, z_0)$ is

_____

© 2011 Cengage Learning. All Rights Reserved. May not be scanned, copied or duplicated, or posted to a publicly accessible website, in whole or in part.

## II. The Angle of Inclination of a Plane (Pages 949–950)

Another use of the gradient $\nabla F(x, y, z)$ is _____

_____.

*What you should learn*
How to find the angle of inclination of a plane in space

The **angle of inclination** of a plane is defined to be _____

_____

_____. The angle of inclination of a horizontal

plane is defined to be _____. Because the vector **k** is

normal to the $xy$-plane, you can use the formula for the cosine of

the angle between two planes to conclude that the angle of

inclination of a plane with normal vector **n** is given by _____

_____.

## III. A Comparison of the Gradients $\nabla f(x, y)$ and $\nabla F(x, y, z)$ (Page 950)

*What you should learn*
How to compare the gradients $\nabla f(x, y)$ and $\nabla F(x, y, z)$

If $F$ is differentiable at $(x_0, y_0, z_0)$ and $\nabla F(x_0, y_0, z_0) \neq \mathbf{0}$, then

$\nabla F(x_0, y_0, z_0)$ is _____ to the level surface

through $(x_0, y_0, z_0)$.

When working with the gradients $\nabla f(x, y)$ and $\nabla F(x, y, z)$, be sure to

remember that $\nabla f(x, y)$ is a vector in _____ and

$\nabla F(x, y, z)$ is a vector in _____.

**Homework Assignment**

Page(s)

Exercises

© 2011 Cengage Learning. All Rights Reserved. May not be scanned, copied or duplicated, or posted to a publicly accessible website, in whole or in part.

## Section 13.8   Extrema of Functions of Two Variables

Course Number

Instructor

Date

**Objective:**   In this lesson you learned how to find absolute and
relative extrema.

**I. Absolute Extrema and Relative Extrema**  (Pages 954–956)

Let $f$ be a continuous function of two variables $x$ and $y$ defined
on a closed bounded region $R$ in the $xy$-plane. The Extreme
Value Theorem states that _____

_____

_____.

Let $f$ be a function defined on a region $R$ containing $(x_0, y_0)$. The
function $f$ has a **relative maximum** at $(x_0, y_0)$ if _____

_____.

The function $f$ has a **relative minimum** $(x_0, y_0)$ if _____

_____.

To say that $f$ has a relative maximum at $(x_0, y_0)$ means that the
point $(x_0, y_0, z_0)$ is _____

_____.

Let $f$ be defined on an open region $R$ containing $(x_0, y_0)$. The
point $(x_0, y_0)$ is a **critical point** of $f$ if one of the following is true.

1.

2.

If $f$ has a relative extremum at $(x_0, y_0)$ on an open region $R$, then

$(x_0, y_0)$ is a _____.

**Example 1:**  Find the relative extrema of
$$f(x, y) = 3x^2 + 2y^2 - 36x + 24y - 9.$$

What you should learn<br>How to find absolute and relative extrema of a function of two variables

© 2011 Cengage Learning. All Rights Reserved. May not be scanned, copied or duplicated, or posted to a publicly accessible website, in whole or in part.

## II.  The Second Partials Test  (Pages 957–959)

The critical points of a function of two variables do not always

yield relative maximum or relative minima. Some critical points

yield _____, which are neither

relative maxima nor relative minima.

*What you should learn*
How to use the Second
Partials Test to find
relative extrema of a
function of two variables

For the Second-Partials Test for Relative Extrema, let $f$ have
continuous second partial derivatives on an open region
containing $(a, b)$ for which $f_x(a, b) = 0$ and $f_y(a, b) = 0$. To test
for relative extrema of $f$, consider the quantity

$$d = f_{xx}(a, b)f_{yy}(a, b) - [f_{xy}(a, b)]^2 .$$

1.  If $d > 0$ and $f_{xx}(a, b) > 0$, then $f$ has _____

    _____.

2.  If $d > 0$ and $f_{xx}(a, b) < 0$, then $f$ has _____

    _____.

3.  If $d < 0$, then _____.

4.  If $d = 0$, then _____.

**Homework Assignment**

Page(s)

Exercises

© 2011 Cengage Learning. All Rights Reserved. May not be scanned, copied or duplicated, or posted to a publicly accessible website, in whole or in part.

## Section 13.9   Applications of Extrema of Functions of Two Variables

**Objective:**    In this lesson you learned how to solve an optimization problem and how to use the method of least squares.

Course Number

Instructor

Date

**I. Applied Optimization Problems**  (Pages 962–963)

Give an example of a real-life situation in which extrema of functions of two variables play a role.

*What you should learn*
How to solve optimization problems involving functions of several variables

Describe the process used to optimize the function of two or more variables.

In many applied problems, the domain of the function to be optimized is a closed bounded region. To find minimum or maximum points, you must not only test critical points, but also

_____

_____ .

**II. The Method of Least Squares**  (Pages 964–966)

In constructing a model to represent a particular phenomenon, the goals are _____ .

*What you should learn*
How to use the method of least squares

As a measure of how well the model $y = f(x)$ fits the collection of points $\{(x_1, y_1), (x_2, y_2), \ldots, (x_n, y_n)\}$, _____

_____

_____ . This sum is called the

_____ and is given by

© 2011 Cengage Learning. All Rights Reserved. May not be scanned, copied or duplicated, or posted to a publicly accessible website, in whole or in part.

_____. Graphically, $S$ can be

interpreted as _____

_____

_____. If the model is a perfect fit, then

$S =$ _____. However, when a perfect fit is not feasible,

you can settle for a model that _____.

The linear model that minimizes $S$ is called _____

_____.

The **least squares regression line** for $\{(x_1, y_1), (x_2, y_2), \ldots,$

$(x_n, y_n)\}$ is given by _____, where

$$a = \frac{n\sum_{i=1}^{n} x_i y_i - \sum_{i=1}^{n} x_i \sum_{i=1}^{n} y_i}{n\sum_{i=1}^{n} x_i^2 - \left(\sum_{i=1}^{n} x_i\right)^2} \quad \text{and} \quad b = \frac{1}{n}\left(\sum_{i=1}^{n} y_i - a\sum_{i=1}^{n} x_i\right).$$

**Example 1:**   Find the least squares regression line for the data
in the table.

| $x$ | 1 | 3 | 4 | 8 | 11 | 12 |
|-----|----|----|----|----|----|----|
| $y$ | 16 | 21 | 24 | 27 | 29 | 33 |

**Homework Assignment**

Page(s)

Exercises

© 2011 Cengage Learning. All Rights Reserved. May not be scanned, copied or duplicated, or posted to a publicly accessible website, in whole or in part.

| Course Number |
| Instructor |
| Date |

## Section 13.10    Lagrange Multipliers

**Objective:**    In this lesson you learned how to solve a constrained
optimization problem using a Lagrange multiplier.

**I. Lagrange Multipliers** (Pages 970–971)

The _____ offers a way

to solve constrained optimization problems.

| *What you should learn* |
| How to understand the Method of Lagrange Multipliers |

Let $f$ and $g$ have continuous first partial derivatives such that $f$ has

an extremum at a point $(x_0, y_0)$ on the smooth constraint curve

$g(x, y) = c$. Lagrange's Theorem states that if $\nabla g(x_0, y_0) \neq \mathbf{0}$, then

there is a real number $\lambda$ such that _____.

The scalar $\lambda$, the lowercase Greek letter lambda, is called a

_____.

Let $f$ and $g$ satisfy the hypothesis of Lagrange's Theorem, and let

$f$ have a minimum or maximum subject to the constraint

$g(x, y) = c$. To find the minimum or maximum of $f$, use the

following steps.

1.

2.

© 2011 Cengage Learning. All Rights Reserved. May not be scanned, copied or duplicated, or posted to a publicly accessible website, in whole or in part.

## II. Constrained Optimization Problems  (Pages 972–974)

Economists call the Lagrange multiplier obtained in a production

function the _____,

which tells the number of additional units of product that can be

produced if one additional dollar is spent on production.

> ***What you should learn***
> How to use Lagrange
> multipliers to solve
> constrained optimization
> problems

## III. The Method of Lagrange Multipliers with Two Constraints  (Page 975)

For an optimization problem involving two constraint functions

$g$ and $h$, you need to introduce _____

_____, and then solve the equation

_____.

> ***What you should learn***
> How to use the Method
> of Lagrange Multipliers
> with two constraints

**Homework Assignment**

Page(s)

Exercises

© 2011 Cengage Learning. All Rights Reserved. May not be scanned, copied or duplicated, or posted to a publicly accessible website, in whole or in part.

# Chapter 14    Multiple Integration

## Section 14.1    Iterated Integrals and Area in the Plane

Course Number

Instructor

Date

**Objective:** In this lesson you learned how to evaluate an iterated integral and find the area of a plane region.

### I. Iterated Integrals  (Pages 984–985)

To extend definite integrals to functions of several variables, you can apply the Fundamental Theorem of Calculus to one variable while _____.

An "integral of an integral" is called a(n) _____.
The _____ limits of integration can be variable with respect to the outer variable of integration. The _____ limits of integration must be constant with respect to both variables of integration. The limits of integration for an iterated integral identify two sets of boundary intervals for the variables, which determine the _____ _____ of the iterated integral.

**Example 1:**   Evaluate $\int_{-3}^{0} \int_{0}^{y} (6x - 2y)\, dx\, dy$ .

> **What you should learn**
> How to evaluate an iterated integral

### II. Area of a Plane Region  (Pages 986–989)

One of the applications of the iterated integral is _____ _____ .

When setting up a double integral to find the area of a region in a plane, placing a representative rectangle in the region R helps determine both _____.
A vertical rectangle implies the order _____,
with the inside limits corresponding to the _____
_____ . This type of region is

> **What you should learn**
> How to use an iterated integral to find the area of a plane region

© 2011 Cengage Learning. All Rights Reserved. May not be scanned, copied or duplicated, or posted to a publicly accessible website, in whole or in part.

called _____, because the

outside limits of integration represent the _____

_____. Similarly, a horizontal

rectangle implies the order _____, with the

inside limits corresponding to the _____

_____. This type of region is called _____

_____, because the outside limits represent

the _____.

**Example 2:**    Use a double integral to find the area of a
rectangular region for which the bounds for $x$ are
$-6 \le x \le 1$ and the bound for $y$ are $-3 \le y \le 8$.

---

**Homework Assignment**

Page(s)

Exercises

---

© 2011 Cengage Learning. All Rights Reserved. May not be scanned, copied or duplicated, or posted to a publicly accessible website, in whole or in part.

## Section 14.2   Double Integrals and Volume

**Objective:**      In this lesson you learned how to use a double integral to find the volume of a solid region.

Course Number

Instructor

Date

### I. Double Integrals and Volume of a Solid Region
   (Pages 992–994)

If $f$ is defined on a closed, bounded region $R$ in the $xy$-plane, then

the _____ is given by

$$\iint\limits_{R} f(x, y)\, dA = \lim_{\|\Delta\| \to 0} \sum_{i=1}^{n} f(x_i, y_i)\Delta A_i \, ,$$ provided the limit exists. If

the limit exists, then $f$ is _____ over $R$.

A double integral can be used to find the volume of a solid

region that lies between _____

_____.

If $f$ is integrable over a plane region $R$ and $f(x, y) \geq 0$ for all $(x, y)$

in $R$, then the volume of the solid region that lies above $R$ and

below the graph of $f$ is defined as _____.

**What you should learn**
How to use a double
integral to represent the
volume of a solid region

### II. Properties of Double Integrals  (Page 994)

Let $f$ and $g$ be continuous over a closed, bounded plane region $R$,

and let $c$ be a constant.

**What you should learn**
How to use properties of
double integrals

1. $\displaystyle\iint\limits_{R} cf(x, y)\, dA = $ _____ $\displaystyle\iint\limits_{R}$ _____

2. $\displaystyle\iint\limits_{R} [f(x, y) \pm g(x, y)]\, dA = \iint\limits_{R}$ _____ $\displaystyle\iint\limits_{R}$ _____

3. $\displaystyle\iint\limits_{R} f(x, y)\, dA \geq 0$ , if _____

© 2011 Cengage Learning. All Rights Reserved. May not be scanned, copied or duplicated, or posted to a publicly accessible website, in whole or in part.

4. $\displaystyle\iint\limits_{R} f(x,y)\,dA \geq \iint\limits_{R} g(x,y)\,dA$, if _____

5. $\displaystyle\iint\limits_{R} f(x,y)\,dA = \iint\limits_{R_1} f(x,y)\,dA + \iint\limits_{R_2} f(x,y)\,dA$, where $R$ is _____

_____ .

### III.  Evaluation of Double Integrals  (Pages 995–999)

Normally, the first step in evaluating a double integral is _____

_____ .

> **What you should learn**
> How to evaluate a double integral as an iterated integral

Explain the meaning of Fubini's Theorem.

In your own words, explain how to find the volume of a solid.

### IV.  Average Value of a Function  (Pages 999–1000)

If $f$ is integrable over the plane region $R$, then the _____

_____ is $\dfrac{1}{A}\displaystyle\iint\limits_{R} f(x,y)\,dA$, where $A$ is the

> **What you should learn**
> How to find the average value of a function over a region

area of $R$.

> **Homework Assignment**
>
> Page(s)
>
> Exercises

© 2011 Cengage Learning. All Rights Reserved. May not be scanned, copied or duplicated, or posted to a publicly accessible website, in whole or in part.

## Section 14.3    Change of Variables:  Polar Coordinates

**Objective:**     In this lesson you learned how to write and evaluate
double integrals in polar coordinates.

Course Number

Instructor

Date

**I. Double Integrals in Polar Coordinates**  (Pages 1004–1008)

Some double integrals are much easier to evaluate in _____

_____ than in rectangular form, especially for regions

such as _____ .

*What you should learn*
How to write and
evaluate double integrals
in polar coordinates

A **polar sector** is defined as _____

_____ .

Let $R$ be a plane region consisting of all points $(x, y) =$
$(r\cos\theta, r\sin\theta)$ satisfying the conditions $0 \le g_1(\theta) \le r \le g_2(\theta)$,
$\alpha \le \theta \le \beta$, where $0 \le (\beta - \alpha) \le 2\pi$. If $g_1$ and $g_2$ are continuous
on $[\alpha, \beta]$ and $f$ is continuous on $R$, then

_____ .

If $z = f(x, y)$ is nonnegative on $R$, then the integral

$$\iint\limits_R f(x, y)\, dA = \int_{\alpha}^{\beta} \int_{g_1(\theta)}^{g_2(\theta)} f(r\cos\theta, r\sin\theta) r\, dr\, d\theta \text{ can be}$$

interpreted as the volume of _____

_____ .

© 2011 Cengage Learning. All Rights Reserved. May not be scanned, copied or duplicated, or posted to a publicly accessible website, in whole or in part.

**Additional notes**

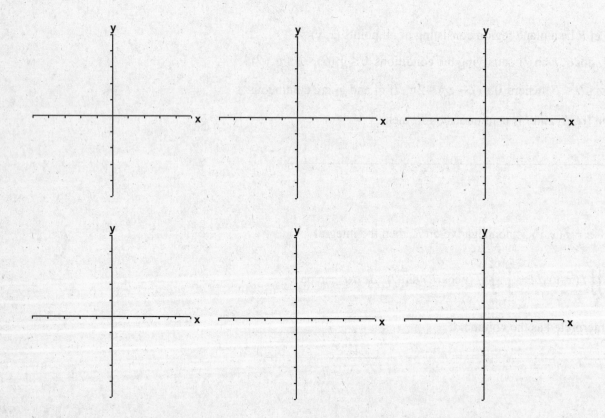

---

**Homework Assignment**

Page(s)

Exercises

---

© 2011 Cengage Learning. All Rights Reserved. May not be scanned, copied or duplicated, or posted to a publicly accessible website, in whole or in part.

## Section 14.4    Center of Mass and Moments of Inertia

**Objective:**    In this lesson you learned how to find the mass of a
planar lamina, the center of mass of a planar lamina, and
moments of inertia using double integrals.

Course Number

Instructor

Date

### I. Mass  (Pages 1012–1013)

If $\rho$ is a continuous density function on the lamina (of variable

density) corresponding to a plane region $R$, then the mass $m$ of

the lamina is given by _____ .

*What you should learn*
How to find the mass of a
planar lamina using a
double integral

For a planar lamina, density is expressed as _____

_____ .

### II. Moments and Center of Mass  (Pages 1014–1015)

Let $\rho$ be a continuous density function on the planar lamina $R$.

The **moments of mass** with respect to the $x$- and $y$-axes are

*What you should learn*
How to find the center of
mass of a planar lamina
using double integrals

$M_x =$ _____ and

$M_y =$ _____ . If $m$ is the mass of the

lamina, then the **center of mass** is _____ .

If $R$ represents a simple plane region rather than a lamina, the

point $(\overline{x}, \overline{y})$ is called the _____ of the

region.

### III. Moments of Inertia  (Pages 1016–1017)

The moments $M_x$ and $M_y$ used in determining the center of mass

of a lamina are sometimes called the _____

about the $x$- and $y$-axes. In each case, the moment is the product

*What you should learn*
How to find moments of
inertia using double
integrals

© 2011 Cengage Learning. All Rights Reserved. May not be scanned, copied or duplicated, or posted to a publicly accessible website, in whole or in part.

of _____. The **second**

**moment,** or the **moment of inertia** of a lamina about a line, is a

measure of _____

_____. These second moments

are denoted $I_x$ and $I_y$, and in each case the moment is the product

of _____.

$I_x =$ _____ and

$I_y =$ _____. The sum of the

moments $I_x$ and $I_y$ is called the _____ and

is denoted by $I_0$.

The moment of inertia $I$ of a revolving lamina can be used to

measure its _____; which is given by

_____, where $\omega$ is the angular speed, in

radians per second, of the planar lamina as it revolves about a

line.

The **radius of gyration** $\overset{=}{r}$ of a revolving mass $m$ with moment

of inertia $I$ is defined to be _____.

---

**Homework Assignment**

Page(s)

Exercises

---

© 2011 Cengage Learning. All Rights Reserved. May not be scanned, copied or duplicated, or posted to a publicly accessible website, in whole or in part.

## Section 14.5    Surface Area

**Objective:**     In this lesson you learned how to use a double integral to find the area of a surface.

Course Number

Instructor

Date

**I. Surface Area** (Pages 1020–1024)

If $f$ and its first partial derivatives are continuous on the closed region $R$ in the $xy$-plane, then the **area of the surface $S$** given by $z = f(x, y)$ over $R$ is given by:

***What you should learn***
How to use a double
integral to find the area
of a surface

List several strategies for performing the often difficult integration involved in finding surface area.

© 2011 Cengage Learning. All Rights Reserved. May not be scanned, copied or duplicated, or posted to a publicly accessible website, in whole or in part.

**Additional notes**

**Homework Assignment**

Page(s)

Exercises

© 2011 Cengage Learning. All Rights Reserved. May not be scanned, copied or duplicated, or posted to a publicly accessible website, in whole or in part.

## Section 14.6   Triple Integrals and Applications

**Objective:**   In this lesson you learned how to use a triple integral to find the volume, center of mass, and moments of inertia of a solid region.

Course Number

Instructor

Date

**I. Triple Integrals**  (Pages 1027–1031)

Consider a function $f$ of three variables that is continuous over a bounded solid region $Q$. Then, encompass $Q$ with a network of boxes and form the _____ consisting of all boxes lying entirely within $Q$. The norm $\|\Delta\|$ of the partition is _____

_____.

If $f$ is continuous over a bounded solid region $Q$, then the **triple integral of $f$ over $Q$** is defined as

_____, provided

the limit exists. The **volume** of the solid region $Q$ is given by

_____.

*What you should learn*
How to use a triple integral to find the volume of a solid region

Let $f$ be continuous on a solid region $Q$ defined by $a \leq x \leq b$, $h_1(x) \leq y \leq h_2(x)$, and $g_1(x, y) \leq z \leq g_2(x, y)$, where $h_1$, $h_2$, $g_1$, and $g_2$ are continuous functions. Then,

_____.

To evaluate a triple iterated integral in the order $dz\,dy\,dx$, _____

_____

_____

_____.

© 2011 Cengage Learning. All Rights Reserved. May not be scanned, copied or duplicated, or posted to a publicly accessible website, in whole or in part.

Describe the process for finding the limits of integration for a triple integral.

## II. Center of Mass and Moments of Inertia
   (Pages 1032–1034)

*What you should learn*
How to find the center of mass and moments of inertia of a solid region

Consider a solid region $Q$ whose density is given by the density function $\rho$. The **center of mass** of a solid region $Q$ of mass $m$ is given by $(\bar{x}, \bar{y}, \bar{z})$ where

$m =$
_____

$M_{yz} =$
_____

$M_{xz} =$
_____

$M_{xy} =$
_____

$\bar{x} =$
_____

$\bar{y} =$
_____

$\bar{z} =$
_____

© 2011 Cengage Learning. All Rights Reserved. May not be scanned, copied or duplicated, or posted to a publicly accessible website, in whole or in part.

The quantities $M_{yz}$, $M_{xz}$, and $M_{xy}$ are called the _____

_____ of the region $Q$ about the $yz$-, $xz$-, and $xy$-

planes, respectively. The first moments for solid regions are

taken about a plane, whereas the second moments for solids are

taken about a _____. The **second moments**

(or **moments of inertia**) about the $x$-, $y$-, and $z$-axes are as

follows.

Moment of inertia about the $x$-axis:  $I_x =$

_____

Moment of inertia about the $y$-axis:  $I_y =$

_____

Moment of inertia about the $z$-axis:  $I_z =$

_____


For problems requiring the calculation of all three moments,
considerable effort can be saved by applying the additive
property of triple integrals and writing

_____

where

$I_{xy} =$

_____

$I_{xz} =$

_____

$I_{yz} =$

_____

© 2011 Cengage Learning. All Rights Reserved. May not be scanned, copied or duplicated, or posted to a publicly accessible website, in whole or in part.

**Additional notes**

Larson/Edwards  **Calculus: Early Transcendental Functions 5e**  Notetaking Guide

**Homework Assignment**

Page(s)

Exercises

© 2011 Cengage Learning. All Rights Reserved. May not be scanned, copied or duplicated, or posted to a publicly accessible website, in whole or in part.

## Section 14.7   Triple Integrals in Cylindrical and Spherical Coordinates

Course Number

Instructor

Date

**Objective:**   In this lesson you learned how to write and evaluate triple integrals in cylindrical and spherical coordinates.

### I. Triple Integrals in Spherical Coordinates
   (Pages 1038–1040)

*What you should learn*
How to write and evaluate a triple integral in cylindrical coordinates

The rectangular conversion equations for cylindrical coordinates are $x =$ _____, $y =$ _____, and $z =$ _____.

If $f$ is a continuous function on the solid $Q$, the iterated form of the triple integral in cylindrical form is

_____.

To visualize a particular order of integration, it helps to view the iterated integral in terms of _____
_____.

For instance, in the order $dr\ d\theta\ dz$, the first integration occurs

_____
_____
_____.

### II. Triple Integrals in Spherical Coordinates
   (Pages 1041–1042)

*What you should learn*
How to write and evaluate a triple integral in spherical coordinates

The rectangular conversion equations for spherical coordinates are $x =$ _____, $y =$ _____, and $z =$ _____.

The triple integral in spherical coordinates for a continuous function $f$ defined on the solid region $Q$ is given by

_____

© 2011 Cengage Learning. All Rights Reserved. May not be scanned, copied or duplicated, or posted to a publicly accessible website, in whole or in part.

As with cylindrical coordinates, you can visualize a particular

order of integration by _____

_____

_____.


**Additional notes**

---

**Homework Assignment**

Page(s)

Exercises

© 2011 Cengage Learning. All Rights Reserved. May not be scanned, copied or duplicated, or posted to a publicly accessible website, in whole or in part.

## Section 14.8    Change of Variables: Jacobians

**Objective:**    In this lesson you learned how to use a Jacobian to change variables in a double integral.

Course Number

Instructor

Date

**I. Jacobians**  (Pages 1045–1046)

If $x = g(u,v)$ and $y = h(u,v)$, then the **Jacobian** of $x$ and $y$ with respect to $u$ and $v$, denoted by $\partial(x,y) / \partial(u,v)$, is

_____ .

***What you should learn***
How to understand the concept of a Jacobian

In general, a change of variables is given by a one-to-one transformation $T$ from a region $S$ in the $uv$-plane to a region $R$ in the $xy$-plane, to be given by _____ _____, where $g$ and $h$ have continuous first partial derivatives in the region $S$. In most cases, you are hunting for a transformation in which _____ _____ .

**II. Change of Variables for Double Integrals**
     (Pages 1047–1049)

Let $R$ be a vertically or horizontally simple region in the $xy$-plane, and let $S$ be a vertically or horizontally simple region in the $uv$-plane. Let $T$ from $S$ to $R$ be given by $T(u, v) = (x, y) = (g(u, v), h(u, v))$, where $g$ and $h$ have continuous first partial derivatives. Assume that $T$ is one-to-one except possibly on the boundary of $S$. If $f$ is continuous on $R$, and $\partial(x,y) / \partial(u,v)$ is nonzero on $S$, then

***What you should learn***
How to use a Jacobian to change variables in a double integral

_____

© 2011 Cengage Learning. All Rights Reserved. May not be scanned, copied or duplicated, or posted to a publicly accessible website, in whole or in part.

**Additional notes**

**Homework Assignment**

Page(s)

Exercises

© 2011 Cengage Learning. All Rights Reserved. May not be scanned, copied or duplicated, or posted to a publicly accessible website, in whole or in part.

# Chapter 15    Vector Analysis

## Section 15.1    Vector Fields

**Objective:** In this lesson you learned how to sketch a vector field, determine whether a vector field is conservative, find a potential function, find curl, and find divergence.

Course Number

Instructor

Date

### I. Vector Fields  (Pages 1058–1061)

A **vector field over a plane region $R$** is _____

_____.

*What you should learn*
How to understand the
concept of a vector field

A **vector field over a solid region $Q$ in space** is _____

_____.

A vector field $\mathbf{F}(x, y, z) = M(x, y, z)\mathbf{i} + N(x, y, z)\mathbf{j} + P(x, y, z)\mathbf{k}$ is

**continuous** at a point if and only if _____

_____.

List some common physical examples of vector fields and give a brief description of each.

Let $\mathbf{r}(t) = x(t)\mathbf{i} + y(t)\mathbf{j} + z(t)\mathbf{k}$ be a position vector. The vector

field $\mathbf{F}$ is **an inverse square field** if

_____, where $k$ is a real number

and $\mathbf{u} = \mathbf{r}/\|\mathbf{r}\|$ is a unit vector in the direction of $\mathbf{r}$.

© 2011 Cengage Learning. All Rights Reserved. May not be scanned, copied or duplicated, or posted to a publicly accessible website, in whole or in part.

Because vector fields consist of infinitely many vectors, it is not possible to create a sketch of the entire field. Instead, when you sketch a vector field, your goal is to _____
_____.

## II. Conservative Vector Fields  (Pages 1061–1063)

The vector field **F** is called **conservative** if _____
_____. The
function *f* is called the _____ for **F**.

Let *M* and *N* have continuous first partial derivatives on an open disk *R*. The vector field given by $\mathbf{F}(x, y) = M\mathbf{i} + N\mathbf{j}$ is

conservative if and only if _____.

*What you should learn*
How to determine whether a vector field is conservative

## III. Curl of a Vector Field  (Pages 1064–1065)

The **curl** of a vector field $\mathbf{F}(x, y, z) = M\mathbf{i} + N\mathbf{j} + P\mathbf{k}$ is


_____.



If curl **F** = **0**, then **F** is said to be _____.

The cross product notation use for curl comes from viewing the gradient $\nabla f$ as the result of the _____
_____ acting on the function *f*.

The primary use of curl is in a test for conservative vector fields in space. The test states _____
_____
_____.

*What you should learn*
How to find the curl of a vector field

© 2011 Cengage Learning. All Rights Reserved. May not be scanned, copied or duplicated, or posted to a publicly accessible website, in whole or in part.

## IV. Divergence of a Vector Field (Page 1066)

*What you should learn*
How to find the
divergence of a vector
field

The curl of a vector field **F** is itself _____.

Another important function defined on a vector field is

**divergence,** which is _____.

The **divergence** of **F**$(x, y) = M\mathbf{i} + N\mathbf{j}$ is

_____.

The **divergence** of **F**$(x, y, z) = M\mathbf{i} + N\mathbf{j} + P\mathbf{k}$ is

_____.

If div **F** = 0, then **F** is said to be _____.

Divergence can be viewed as _____

_____

_____

_____.

If **F**$(x, y, z) = M\mathbf{i} + N\mathbf{j} + P\mathbf{k}$ is a vector field and $M, N,$ and $P$ have

continuous second partial derivatives, then _____.

© 2011 Cengage Learning. All Rights Reserved. May not be scanned, copied or duplicated, or posted to a publicly accessible website, in whole or in part.

**Additional notes**

**Homework Assignment**

Page(s)

Exercises

© 2011 Cengage Learning. All Rights Reserved. May not be scanned, copied or duplicated, or posted to a publicly accessible website, in whole or in part.

## Section 15.2    Line Integrals

**Objective:**       In this lesson you learned how to find a piecewise
smooth parametrization, and write and evaluate a line
integral.

Course Number

Instructor

Date

### I. Piecewise Smooth Curves  (Page 1069)

A plane curve $C$ given by $\mathbf{r}(t) = x(t)\mathbf{i} + y(t)\mathbf{j}$, $a \le t \le b$, is

**smooth** if _____

_____. A space

curve C given by $\mathbf{r}(t) = x(t)\mathbf{i} + y(t)\mathbf{j} + z(t)\mathbf{k}$, $a \le t \le b$, is

**smooth** if _____

_____. A curve $C$ is

**piecewise smooth** if _____

_____

_____.

*What you should learn*
How to understand and
use the concept of a
piecewise smooth curve

### II. Line Integrals  (Pages 1070–1073)

If $f$ is defined in a region containing a smooth curve $C$ of finite
length, then the **line integral of $f$ along $C$** is given by

for a plane

_____

or by

_____

for space, provided this limit exists.

Let $f$ be continuous in a region containing a smooth curve $C$. If $C$
is given by $\mathbf{r}(t) = x(t)\mathbf{i} + y(t)\mathbf{j}$, where $a \le t \le b$, then

_____

If $C$ is given by $\mathbf{r}(t) = x(t)\mathbf{i} + y(t)\mathbf{j} + z(t)\mathbf{k}$, where $a \le t \le b$, then

_____

*What you should learn*
How to write and
evaluate a line integral

© 2011 Cengage Learning. All Rights Reserved. May not be scanned, copied or duplicated, or posted to a publicly accessible website, in whole or in part.

If $f(x, y, z) = 1$, the line integral gives _____

_____.

## III.  Line Integrals of Vector Fields  (Pages 1074–1076)

Let **F** be a continuous vector field defined on a smooth curve $C$

given by $\mathbf{r}(t)$, $a \le t \le b$. The **line integral** of **F** on $C$ is given by

_____

*What you should learn*
How to write and
evaluate a line integral of
a vector field

## IV.  Line Integrals in Differential Form  (Pages 1077–1078)

If **F** is a vector field of the form $\mathbf{F}(x, y) = M\mathbf{i} + N\mathbf{j}$, and $C$ is given

by $\mathbf{r}(t) = x(t)\mathbf{i} + y(t)\mathbf{j}$, then $\mathbf{F} \cdot d\mathbf{r}$ is often written in **differential**

**form** as _____.

*What you should learn*
How to write and
evaluate a line integral in
differential form

**Homework Assignment**

Page(s)

Exercises

© 2011 Cengage Learning. All Rights Reserved. May not be scanned, copied or duplicated, or posted to a publicly accessible website, in whole or in part.

## Section 15.3   Conservative Vector Fields and Independence of Path

Course Number

Instructor

Date

**Objective:**   In this lesson you learned how to use the Fundamental Theorem of Line Integrals, independence of path, and conservation of energy.

### I. Fundamental Theorem of Line Integrals
   (Pages 1083–1085)

*What you should learn*
How to understand and use the Fundamental Theorem of Line Integrals

Let $C$ be a piecewise smooth curve lying in an open region $R$ and given by $\mathbf{r}(t) = x(t)\mathbf{i} + y(t)\mathbf{j}$, $a \leq t \leq b$. The **Fundamental Theorem of Line Integrals** states that if $\mathbf{F}(x, y) = M\mathbf{i} + N\mathbf{j}$ is conservative in $R$, and $M$ and $N$ are continuous in $R$, then

_____

where $f$ is a potential function of $\mathbf{F}$. That is, $\mathbf{F}(x, y) = \nabla f(x, y)$.

The Fundamental Theorem of Line Integrals states that _____

_____

_____

_____.

### II. Independence of Path   (Pages 1086–1088)

*What you should learn*
How to understand the concept of independence of path

Saying that the line integral $\displaystyle\int_C \mathbf{F} \cdot d\mathbf{r}$ is **independent of path**

means that _____

_____

_____.

If $\mathbf{F}$ is continuous on an open connected region, then the line integral $\displaystyle\int_C \mathbf{F} \cdot d\mathbf{r}$ is independent of path if and only if _____

_____.

© 2011 Cengage Learning. All Rights Reserved. May not be scanned, copied or duplicated, or posted to a publicly accessible website, in whole or in part.

A curve $C$ given by $\mathbf{r}(t)$ for $a \leq t \leq b$ is **closed** if

_____.

Let $\mathbf{F}(x, y, z) = M\mathbf{i} + N\mathbf{j} + P\mathbf{k}$ have continuous first partial derivatives in an open connected region $R$, and let $C$ be a piecewise smooth curve in $R$. The following conditions are equivalent.

1.

2.

3.

### III.  Conservation of Energy  (Page 1089)

State the Law of Conservation of Energy.

*What you should learn*
How to understand the concept of conservation of energy

The **kinetic energy** of a particle of mass $m$ and speed $v$ is

_____.

The **potential energy** $p$ of a particle at point $(x, y, z)$ in a conservative vector field $\mathbf{F}$ is defined as _____

_____, where $f$ is the potential function for $\mathbf{F}$.

**Homework Assignment**

Page(s)

Exercises

© 2011 Cengage Learning. All Rights Reserved. May not be scanned, copied or duplicated, or posted to a publicly accessible website, in whole or in part.

## Section 15.4   Green's Theorem

**Objective:**     In this lesson you learned how to evaluate a line integral
using Green's Theorem.

Course Number

Instructor

Date

**I.  Green's Theorem**  (Pages 1093–1098)

A curve $C$ given by $\mathbf{r}(t) = x(t)\mathbf{i} + y(t)\mathbf{j}$, where $a \le t \le b$, is

**simple** if _____ —that is,

$\mathbf{r}(c) \ne \mathbf{r}(d)$ for all $c$ and $d$ in the open interval $(a, b)$. A plane

region $R$ is **simply connected** if _____

_____.

***What you should learn***
How to use Green's
Theorem to evaluate a
line integral

Let $R$ be a simply connected region with a piecewise smooth

boundary $C$, oriented counterclockwise (that is, $C$ is traversed

once so that the region $R$ always lies to the left). Then **Green's**

**Theorem** states that if $M$ and $N$ have continuous first partial

derivatives in an open region containing $R$, then

_____.

**Line Integral for Area**

If $R$ is a plane region bounded by a piecewise smooth simple

closed curve $C$, oriented counterclockwise, then the area of $R$ is

given by _____.

**II.  Alternative Forms of Green's Theorem**
     (Pages 1098–1099)

With appropriate condition on $\mathbf{F}$, $C$, and $R$, you can write

Green's Theorem in the following vector form

***What you should learn***
How to use alternative
forms of Green's
Theorem

© 2011 Cengage Learning. All Rights Reserved. May not be scanned, copied or duplicated, or posted to a publicly accessible website, in whole or in part.

For the second vector form of Green's Theorem, assume the

same conditions for **F**, $C$, and $R$. Using the arc length parameter

$s$ for $C$, you have _____. So, a unit

tangent vector **T** to curve $C$ is given by

_____. The outward unit

normal vector **N** can then be written as

_____. The second alternative form

of Green's Theorem is given by

_____.

---

**Homework Assignment**

Page(s)

Exercises

---

© 2011 Cengage Learning. All Rights Reserved. May not be scanned, copied or duplicated, or posted to a publicly accessible website, in whole or in part.

## Section 15.5 Parametric Surfaces

**Objective:** In this lesson you learned how to sketch a parametric surface, find a set of parametric equations to represent a surface, find a normal vector, find a tangent plane, and find the area of a parametric surface.

Course Number

Instructor

Date

### I. Parametric Surfaces (Pages 1102–1103)

Let $x$, $y$, and $z$ be functions of $u$ and $v$ that are continuous on a domain $D$ in the $uv$-plane. The set of points $(x, y, z)$ given by

$\mathbf{r}(u,v) = x(u,v)\mathbf{i} + y(u,v)\mathbf{j} + z(u,v)\mathbf{k}$ is called a _____

_____. The equations $x = x(u, v)$, $y = y(u, v)$, and

$z = z(u, v)$ are the _____ for the

surface.

If $S$ is a parametric surface given by the vector-valued function $\mathbf{r}$,

then $S$ is traced out by _____

_____.

*What you should learn*
How to understand the definition of a parametric surface, and sketch the surface

### II. Finding Parametric Equations for Surfaces (Page 1104)

Writing a set of parametric equations for a given surface is

generally more difficult than identifying the surface described by

a given set of parametric equations. One type of surface for

which this problem is straightforward, however is the surface

given by $z = f(x, y)$. You can parameterize such a surface as

_____.

*What you should learn*
How to find a set of parametric equations to represent a surface

### III. Normal Vectors and Tangent Planes (Pages 1105–1106)

Let $S$ be a smooth parametric surface

$\mathbf{r}(u,v) = x(u,v)\mathbf{i} + y(u,v)\mathbf{j} + z(u,v)\mathbf{k}$ defined over an open region

$D$ in the $uv$-plane. Let $(u_0, v_0)$ be a point in $D$. A normal vector at

the point $(x_0, y_0, z_0) = (x(u_0, v_0), y(u_0, v_0), z(u_0, v_0))$ is given by

*What you should learn*
How to find a normal vector and a tangent plane to a parametric surface

© 2011 Cengage Learning. All Rights Reserved. May not be scanned, copied or duplicated, or posted to a publicly accessible website, in whole or in part.

_____.

**IV.  Area of a Parametric Surface**  (Pages 1106–1108)

Let $S$ be a smooth parametric surface

$\mathbf{r}(u,v) = x(u,v)\mathbf{i} + y(u,v)\mathbf{j} + z(u,v)\mathbf{k}$  defined over an open region

$D$ in the $uv$-plane. If each point on the surface $S$ corresponds to

exactly one point in the domain $D$, then the **surface area** $S$ is

given by _____,

where $\mathbf{r}_u = $ _____       and

           _____

$\mathbf{r}_v = $ _____.

| What you should learn |
| --- |
| How to find the area of a parametric surface |

**Homework Assignment**

Page(s)

Exercises

© 2011 Cengage Learning. All Rights Reserved. May not be scanned, copied or duplicated, or posted to a publicly accessible website, in whole or in part.

## Section 15.6 Surface Integrals

**Objective:** In this lesson you learned how to evaluate a surface integral, determine the orientation of a surface, and evaluate a flux integral.

Course Number

Instructor

Date

### I. Surface Integrals (Pages 1112–1115)

Let $S$ be a surface with equation $z = g(x, y)$ and let $R$ be its projection onto the $xy$-plane. If $g$, $g_x$, and $g_y$ are continuous on $R$ and $f$ is continuous on $S$, then the **surface integral of $f$ over $S$** is

_____

***What you should learn***
How to evaluate a surface integral as a double integral

### II. Parametric Surfaces and Surface Integrals (Page 1116)

For a surface $S$ given by the vector-valued function $\mathbf{r}(u,v) = x(u,v)\mathbf{i} + y(u,v)\mathbf{j} + z(u,v)\mathbf{k}$ defined over a region $D$ in the $uv$-plane, you can show that the surface integral of $f(x, y, z)$ over $S$ is given by

_____

***What you should learn***
How to evaluate a surface integral for a parametric surface

### III. Orientation of a Surface (Page 1117)

Unit normal vectors are used to _____
_____. A surface is called **orientable** if
_____
_____
_____.

If this is possible, $S$ is called _____.

***What you should learn***
How to determine the orientation of a surface

### IV. Flux Integrals (Pages 1118–1121)

Suppose an oriented surface $S$ is submerged in a fluid having a continuous velocity field $\mathbf{F}$. Let $\Delta S$ be the area of a small patch of the surface $S$ over which $\mathbf{F}$ is nearly constant. Then the amount of fluid crossing this region per unit time is

***What you should learn***
How to understand the concept of a flux integral

© 2011 Cengage Learning. All Rights Reserved. May not be scanned, copied or duplicated, or posted to a publicly accessible website, in whole or in part.

approximated by_____

_____. Consequently, the volume of fluid

crossing the surface $S$ per unit time is called _____

_____.

Let $\mathbf{F}(x, y, z) = M\mathbf{i} + N\mathbf{j} + P\mathbf{k}$, where $M$, $N$, and $P$ have

continuous first partial derivatives on the surface $S$ oriented by a

unit normal vector $\mathbf{N}$. The **flux integral of F across** $S$ is given

by _____.

Let $S$ be an oriented surface given by $z = g(x, y)$ and let $R$ be its

projection onto the $xy$-plane. If the surface is oriented upward,

$$\iint_S \mathbf{F} \cdot \mathbf{N} \, dS = \underline{\hspace{6cm}}. \text{ If}$$

the surface is oriented downward, $\iint_S \mathbf{F} \cdot \mathbf{N} \, dS =$

_____.

---

**Homework Assignment**

Page(s)

Exercises

---

© 2011 Cengage Learning. All Rights Reserved. May not be scanned, copied or duplicated, or posted to a publicly accessible website, in whole or in part.

## Section 15.7    Divergence Theorem

**Objective:**     In this lesson you learned how to use the Divergence
Theorem and how to use it to calculate flux.

Course Number

Instructor

Date

**I. Divergence Theorem** (Pages 1124–1128)

The **Divergence Theorem** gives the relationship between _____
_____
_____.

***What you should learn***
How to understand and
use the Divergence
Theorem

In the Divergence Theorem, the surface $S$ is **closed** in the sense

that it _____

_____.

Let $Q$ be a solid region bounded by a closed surface $S$ oriented

by a unit normal vector directed outward from $Q$. The

**Divergence Theorem** states that if $\mathbf{F}$ is a vector field whose

component functions have continuous first partial derivatives in

$Q$, then _____.

**II. Flux and the Divergence Theorem** (Pages 1129–1130)

Consider the two sides of the equation

$$\iint_S \mathbf{F} \cdot \mathbf{N}\, dS = \iiint_Q \operatorname{div} \mathbf{F}\, dV$$ . The flux integral on the left

determines _____

_____. This can be approximated by

_____

The triple integral on the right measures _____

_____

_____

***What you should learn***
How to use the
Divergence Theorem to
calculate flux

© 2011 Cengage Learning. All Rights Reserved. May not be scanned, copied or duplicated, or posted to a publicly accessible website, in whole or in part.

The point $(x_0, y_0, z_0)$ in a vector field is classified as a **source** if
_____; a **sink** if _____, or
**incompressible** if _____.

In hydrodynamics, a *source* is a point at which _____

_____

_____. A *sink* is a point at which _____

_____.

<br>
<br>
<br>

---

**Homework Assignment**

Page(s)

Exercises

---

© 2011 Cengage Learning. All Rights Reserved. May not be scanned, copied or duplicated, or posted to a publicly accessible website, in whole or in part.

## Section 15.8 Stokes's Theorem

**Objective:** In this lesson you learned how to use Stokes's Theorem to evaluate a line integral or a surface integral and how to use curl to analyze the motion of a rotating liquid.

Course Number

Instructor

Date

### I. Stokes's Theorem (Pages 1132–1134)

**Stokes's Theorem** gives the relationship between _____

_____

_____.

*What you should learn*
How to understand and use Stokes's Theorem

The positive direction along $C$ is _____ relative to the normal vector **N**. That is, if you imagine grasping the normal vector **N** with your right hand, with your thumb pointing in the direction of **N**, your fingers will point _____

_____.

Let $S$ be an oriented surface with unit normal vector **N**, bounded by a piecewise smooth simple closed curve $C$ with a positive orientation. **Stokes's Theorem** states that if **F** is a vector field whose component functions have continuous first partial derivatives on an open region containing $S$ and $C$, then

_____.

### II. Physical Interpretation of Curl (Pages 1135–1136)

curl $\mathbf{F}(x, y, z) \cdot \mathbf{N} =$ _____

*What you should learn*
How to use curl to analyze the motion of a rotating liquid

The rotation of **F** is maximum when _____

_____. Normally, this tendency to rotate will vary from point to point on the surface $S$, and Stokes's Theorem says that the collective measure of this rotational tendency taken over the entire surface $S$ (surface integral) is equal to _____

_____.

© 2011 Cengage Learning. All Rights Reserved. May not be scanned, copied or duplicated, or posted to a publicly accessible website, in whole or in part.

If curl **F** = **0** throughout region $Q$, the rotation of **F** about each

unit normal **N** is _____. That is, **F** is

_____.

<br>

---

**Homework Assignment**

Page(s)

Exercises

© 2011 Cengage Learning. All Rights Reserved. May not be scanned, copied or duplicated, or posted to a publicly accessible website, in whole or in part.

# Chapter 16    Additional Topics in Differential Equations

Course Number

Instructor

Date

## Section 16.1    Exact First-Order Equations

**Objective:** In this lesson you learned how to recognize and solve exact differential equations.

### I. Exact Differential Equations  (Pages 1144–1146)

*What you should learn*
How to solve an exact differential equation

The equation $M(x, y)dx + N(x, y)dy = 0$ is an **exact differential**

**equation** if _____

_____

_____.

The general solution of the equation is _____.

State the **Test for Exactness.**

A _____ is actually

a special type of an exact equation.

**Example 1:**   Test whether the differential equation
$$(5x - x^3 y)dx + \left(y - \tfrac{1}{4}x^4\right)dy = 0 \text{ is exact.}$$

A general solution $f(x, y) = C$ to an exact differential equation

can be found by _____

_____.

© 2011 Cengage Learning. All Rights Reserved. May not be scanned, copied or duplicated, or posted to a publicly accessible website, in whole or in part.

**II. Integrating Factors**  (Pages 1147–1148)

If the differential equation $M(x, y)dx + N(x, y)dy = 0$ is not

exact, it may be possible to make it exact by _____

_____

_____ .

> **What you should learn**
> How to use an integrating factor to make a differential equation exact

Consider the differential equation $M(x, y)dx + N(x, y)dy = 0$. If

$\dfrac{1}{N(x, y)}[M_y(x, y) - N_x(x, y)] = h(x)$ is a function of $x$ alone,

then _____ is an integrating factor. If

$\dfrac{1}{M(x, y)}[N_x(x, y) - M_y(x, y)] = k(y)$ is a function of $y$ alone,

then _____ is an integrating factor.

**Additional notes**

**Homework Assignment**

Page(s)

Exercises

© 2011 Cengage Learning. All Rights Reserved. May not be scanned, copied or duplicated, or posted to a publicly accessible website, in whole or in part.

## Section 16.2    Second-Order Homogeneous Linear Equations

**Objective:**    In this lesson you learned how to solve second-order homogeneous linear differential equations and higher-order homogeneous linear differential equations.

| Course Number |
|---|
| Instructor |
| Date |

### I.  Second-Order Linear Differential Equations
(Pages 1151–1154)

> **What you should learn**
> How to solve a second-order linear differential equation

Let $g_1, g_2, \ldots g_n$ and $f$ be functions of $x$ with a common (interval) domain. An equation of the form

$$y^{(n)} + g_1(x)y^{(n-1)} + g_2(x)y^{(n-2)} + \cdots + g_{n-1}(x)y' + g_n(x)y = f(x)$$

is called a _____ .

If $f(x) = 0$, the equation is _____ ;

otherwise, it is _____ .

The functions $y_1, y_2, \ldots, y_n$ are _____ if

the only solution of the equation $C_1 y_1 + C_2 y_2 + \cdots + C_n y_n = 0$ is

the trivial one, $C_1 = C_2 = \cdots = C_n = 0$. Otherwise, this set of

functions is _____ .

If $y_1$ and $y_2$ are linearly independent solutions of the differential

equation $y'' + ay' + by = 0$, then the general solution is

_____ , where $C_1$ and $C_2$ are constants.

In other words, if you can find two linearly independent

solutions, you can obtain the general solution by _____

_____ .

The **characteristic equation** of the differential equation

$y'' + ay' + by = 0$ is _____ .

The solutions of $y'' + ay' + by = 0$ fall into one of the following
there cases, depending on the solutions of the characteristic
equation, $m^2 + am + b = 0$.

1.

© 2011 Cengage Learning. All Rights Reserved. May not be scanned, copied or duplicated, or posted to a publicly accessible website, in whole or in part.

2.

3.

## II. Higher-Order Linear Differential Equations  (Page 1155)

Describe how to solve higher-order homogeneous linear
differential equations.

*What you should learn*
How to solve a higher-
order linear differential
equation

## III. Application  (Pages 1156–1157)

Describe Hooke's Law.

*What you should learn*
How to use a second-
order linear differential
equation to solve an
applied problem

The equation that describes the undamped motion of a spring is

_____.

---

**Homework Assignment**

Page(s)

Exercises

---

© 2011 Cengage Learning. All Rights Reserved. May not be scanned, copied or duplicated, or posted to a publicly accessible website, in whole or in part.

## Section 16.3    Second-Order Nonhomogeneous Linear Equations

Course Number

Instructor

Date

**Objective:**    In this lesson you learned how to solve second-order nonhomogeneous linear differential equations.

### I. Nonhomogeneous Equations  (Page 1159)

Let $y'' + ay' + by = F(x)$ be a second-order nonhomogeneous linear differential equation. If $y_p$ is a particular solution of this equation and $y_h$ is the general solution of the corresponding homogeneous equation, then _____

is the general solution of the nonhomogeneous equation.

> ***What you should learn***
> How to recognize the general solution of a second-order nonhomogeneous linear differential equation

### II. Method of Undetermined Coefficients  (Pages 1160–1162)

If $F(x)$ in $y'' + ay' + by = F(x)$ consists of sums or products of $x^n$, $e^{mx}$, $\cos \beta x$, or $\sin \beta x$, you can find a particular solution $y_p$ by the method of _____.

Describe how to use this method.

> ***What you should learn***
> How to use the method of undetermined coefficients to solve a second-order nonhomogeneous linear differential equation

### III. Variation of Parameters  (Pages 1163–1164)

Describe the conditions to which the method of variation of parameters is best suited.

> ***What you should learn***
> How to use the method of variation of parameters to solve a second-order nonhomogeneous linear differential equation

© 2011 Cengage Learning. All Rights Reserved. May not be scanned, copied or duplicated, or posted to a publicly accessible website, in whole or in part.

To use the method of variation of parameters to find the general
solution of the equation $y'' + ay' + by = F(x)$, use the following
steps.

1.

2.

3.

4.

**Homework Assignment**

Page(s)

Exercises

© 2011 Cengage Learning. All Rights Reserved. May not be scanned, copied or duplicated, or posted to a publicly accessible website, in whole or in part.

## Section 16.4    Series Solutions of Differential Equations

**Objective:**    In this lesson you learned how to use power series to solve differential equations.

Course Number

Instructor

Date

**I.  Power Series Solution of a Differential Equation**
    (Page 1167–1168)

Recall that a power series represents a function $f$ on _____

_____, and you can successively differentiate

the power series to obtain a series for $f'$, $f''$, and so on.

Describe how to use power series in the solution of a differential equation.

*What you should learn*
How to use a power
series to solve a
differential equation

**II.  Approximation by Taylor Series**  (Page 1169)

What type of series can be used to solve differential equations with initial conditions?

*What you should learn*
How to use a Taylor
series to find the series
solution of a differential
equation

Describe how to use this method.

© 2011 Cengage Learning. All Rights Reserved. May not be scanned, copied or duplicated, or posted to a publicly accessible website, in whole or in part.

**Additional notes**

**Homework Assignment**

Page(s)

Exercises

© 2011 Cengage Learning. All Rights Reserved. May not be scanned, copied or duplicated, or posted to a publicly accessible website, in whole or in part.